The Art of the
Sound Effects Editor

The Art of the Sound Effects Editor

Marvin M. Kerner

Focal Press
Boston London

Focal Press is an imprint of Butterworth–Heinemann

Library of Congress Cataloging-in-Publication Data
Kerner, Marvin M.
    The art of the sound effects editor / Marvin M. Kerner.
        p.   cm.
    Includes index.
    ISBN 0-240-80008-7
    1. Motion pictures—Sound effects. 2. Motion pictures—Editing.
    I. Title.
    TR899.K47   1989
    778.5'344—dc19
                                                                88-18843
                                                                    CIP

British Library Cataloging in Publication Data
Kerner, Marvin M.
    The art of the sound effects editor.
    1. Cinema films. Sound effects. Editing
    I. Title
    778.5'344
    ISBN 0-240-80008-7

Butterworth–Heinemann
313 Washington Street
Newton, MA 02158–1626

10 9 8 7 6 5 4 3 2

Printed in the United States of America

Dedication

On August 8, 1955, I went to work as a bright-eyed apprentice in the sound effects department. Looking around, it was soon apparent that the staff editors were an aging group and if I kept my eyes open and my ears clean I could have a long career at the studio.

As often happens in life, things did not work out that way. More years have passed than I care to count since then. The studio underwent many changes as did the industry itself. We have seen the demise of the big studio and the rise of the independent sound editing company. Over the years one thing has remained: the influence of that handful of *old timers* with whom I was close.

They were a crusty group of often malcontent individuals with one common trait: dedication to quality, creative sound editing. The difference between a good editor and a poor one is not IQ; it is in the caring. Under the big studio system the good editors received no benefit over the others than the personal satisfaction they derived from their work.

To Mike Steinor, Ralph George, Kurt Herrenfeld, and Milo Lowry, thanks for setting a standard that I have attempted to maintain. Especially to John Lipow, my personal tutor, thanks for the standard for which I am still reaching.

Contents

Introduction   **1**
Author's Perspective   1
The Sound Effects Editor   2
Job Responsibilities   2
Sound Editing: A Separate Branch   3
The Art of Sound Effects   4

1   A Brief History of Sound Effects   7
The Foley Stage   7
Looping   8
Magnetic Tape   9
Productivity   9
The Independents   9

2   The Purpose of Sound Effects   11
Simulating Reality   11
Creating Illusion   11
Mood   12
Silence   14

3   **Spotting   17**
    A Word of Caution   17
    Spotting a Movie of the Week   18
    Off Camera Effects   19

4   **Preparation   21**
    Judgment   22
    Kits   26

5   **Dialog Tracks   27**
    Futz   31
    Importance of Saving Production Sound   32

6   **Automated Dialog Replacement (Looping)   33**
    Programming   34
    Recording ADR   36
    Wild Lines   37
    Editing ADR   38
    Actors   39

7   **Selecting the Effects   41**
    Satisfaction with the Effects   41
    Crowds   42
    Horses   42
    Backgrounds   43
    The Fake   43
    Multiple Effects   45

8   **Ordering Material from the Library   47**
    Additional Sources of Sound   48

9   **Cutting   51**
    Non-Sync Effects   52
    Sync Effects   53
    Review   55
    Tricks of the Trade   56
    Conclusion   59

10 **Building   61**
    Summary   65

11 **The Foley Stage   67**
    Separation   71
    Foley Stage versus Library   71

12 Dubbing 73
Editor–Mixer Relationship 73
The Process 75
Preparation 80
Cue Sheets 80
The Change Room 81
Mixing the Good with the Bad 82

13 The Library 83
Origins of Sound Effects Libraries 83
An Important Tool 84
Library Organization 84
Care and Maintenance 89
Library Formats 90

14 The Assistant 91

15 Foreign Language Films 93
Foreign Films from Separation Dubs 93
Foreign Films from Composites 94

16 *Dirty Harry*: A Critique 95

17 Summary 99

Glossary 101

Index 105

Introduction

AUTHOR'S PERSPECTIVE

In 1927 Warner Brothers Studios released *The Jazz Singer*. The sound craze swept the motion picture industry. The Metro-Goldwyn-Mayer (MGM) management asked Ralph George from the electric department to organize a sound effects editing department. Forty-three years later Ralph retired, a crusty veteran of all the great MGM films over the years. It was my pleasure to know and work for Ralph. His intimidating glare taught his staff discipline, and his depth of experience taught his staff to care about sound editing.

I started my career as an apprentice at MGM in August 1955. I worked as an apprentice on *Ben Hur*, an assistant on *Dr. Zhivago*, and an editor on *Ryan's Daughter*, *Shoes of the Fisherman*, and *Grand Prix*. I was also an editor on the "Man from U.N.C.L.E." television series. In 1975 I started my own post-production company, and in the intervening years I have been the sole or supervising editor on approximately 50 TV movies and a similar number of mostly low-budget feature films. A brief list of credits includes such features as *Buster and Billie*, *White Lightning*, *Creature*, *Lust in the Dust*, and *Outlaw Blues*. My TV credits include the eight-hour "The Word" and four-hour shows such as "Rumor of War," "Dempsey," and most recently, "Windmills of the Gods."

I have at times belonged to the Motion Picture Sound Editors, the

1

Academy of Television Arts and Sciences, and the Academy of Motion Picture Arts and Sciences.

THE SOUND EFFECTS EDITOR

The sound effects editor is an important and creative member of the film production team. Unfortunately, there is a lack of available material on the subject of sound effects and its filmic usage. There is also a similar void in the teaching of the subject. Few film school teachers have the necessary expertise to teach the correct usage of sound effects.

For an aspiring producer, director, writer, unit production manager, or film editor, an understanding of how sound effects can be used is a vital part of the education. You cannot consider yourself an expert until you have an understanding of each step, each craft and science required to prepare and shoot a film.

In this text, we shall approach the subject of sound effects editing as one step in your education. The more you know about film, the better opportunity you have of landing a job. Once you have a job, that knowledge will help you become better in your chosen craft.

JOB RESPONSIBILITIES

Film editorial work appeals to many film school graduates for several reasons. Not the least of these is the availability of jobs. That is not to say the jobs are out there waiting for you. They are hard to come by as are jobs in any part of the film industry. The difficulty of finding work right out of school is emphasized by the fact that there are probably ten people for every available job in Hollywood. Nevertheless, editorial work can sometimes offer a relatively easy way into the film industry. As a graduate, you will probably have made one or more films of your own and in the process done some amount of editing. Hopefully, this will include both film and sound editing. The more experience you have, the better your chances of landing a job in the film industry.

The Film Apprentice

An entry-level film apprentice basically does everything the assistant does not want to do. The apprentice carries film. I once worked at Columbia as an apprentice in the film shipping department. The big film cans seen around theaters carrying the release prints are made by a company named Goldberg Bros. With 4000 to 6000 feet of film and metal reels included, these cans are heavy, and can be cumbersome to carry around all day in and out of projection booths. The apprentice makes box labels for the dailies. The apprentice sometimes codes film.

The Assistant

The assistant basically does everything the editor does not want to do and the apprentice does not know how to do. The film assistant codes, builds dailies (i.e., syncing the picture and sound from each day's shooting), breaks down dailies and files them in boxes prepared by the apprentice, pulls scenes for the editor, and refiles trims. The assistant may also have the opportunity of learning something about editing from the editor. Much of the assistant's education comes from simply sitting in various screenings with the producer and/or director at which time editing is discussed. The sound effects assistant also has the responsibility of getting all the material together for the editor. Often the assistant builds tracks (splices into leader by footages marked on them by the editor). The assistant may also do the dubbing cue sheets and perhaps spend some time on the dubbing stage with the film, which can be a wonderful learning experience.

The Editor

The sound effects editor has two responsibilities: (1) to take the production sound track and properly prepare it for dubbing (dialog tracks are discussed in Chapter 5), and (2) to add sound effects to the film. Competency in both areas are prerequisites to the position of sound effects editor.

As a point of information, the terms *sound editor* and *sound effects editor* are interchangeable.

In the modern era of specialization, we are finding, most particularly in big-budget films, a dividing of responsibilities. An editor might be responsible for backgrounds, another for specific effects, a third Foley, and yet another for automated dialog replacement (ADR) (see Chapter 6), not to mention the dialog track editors. There is nothing wrong with this approach assuming that each of the editors is capable of fulfilling any of the jobs at a given time.

SOUND EDITING: A SEPARATE BRANCH

Originally, the position of sound editors in the film production was unclear. The sound department laid some claim to them because they were working in sound. The editorial department laid their claim based on the fact that they worked with film and edited. This lack of specific base holds true even today. In the Academy of Motion Picture Arts and Sciences the sound editors have their own branch. Thus, they do not fit into the sound branch or in the category of film editors.

THE ART OF SOUND EFFECTS

Unlike painting, sculpture, or any of the arts, filmmaking is a team project rather than the product of a single talent. One writer can produce a fine script, one artist can paint the Mona Lisa, but a film is the result of many talents pooled to create the greatest communication vehicle known to humans. It is likely that more people see a television movie in one evening than have ever seen the Mona Lisa. That television show can have a greater impact (admittedly for either good or bad) on more people than the greatest works of art.

I will show later in this text how a telephone bell cut one way rather than another can affect the mood of a scene or establish an actor's demeanor. I will also demonstrate how a single sound effect placed before or after a line of dialog can (and does) affect not only the mood of the dialog but also the meaning of the words.

The art of the sound editor is an integral part of any film. To prove this point, imagine watching a film before the sound editor has worked on it. Watch it with just the production sound track, with all its imperfections, without the necessary looping and sound effects. Consider how convincing a gun battle is with the sound of cap pistols coming from Dirty Harry's gun. Or, consider the reality of a *Star Wars* battle without sound. The car chase in *Bullitt* was surely one of the best ever done, but it would lose much of its impact without the roaring engines and exaggerated skids. Bruce Lee is less menacing when he karate chops the villain into oblivion without the swish, the crack, the grunt, and the body fall—all created by the sound effects editor.

In measuring the relative impact on a film, one should first number the producer, followed in some order by the screenwriter, the director, the actors, the cinematographer, the set designer, the wardrobe designer, and the film editor. Then most certainly follows the sound editor.

The sound editor does not receive the money or the credit that his or her contribution to the film deserves. Current union contracts call for the film editor to get single card credit up front. The sound editor's credit is mired in the end credits somewhere after the producer's secretary and maybe just before the transportation captain. I would like to see a film released without the sound editor's work. It is not my intention to demean the contribution of the secretary or the truck driver. Film is, however, a medium of sight and sound. The sound in any film is the direct result of the sound editor's work and craft . . . in many cases his art.

In order to help the reader develop the clearest understanding of how the sound effects editor works on a film, this book's chapters have been arranged so that the topics discussed roughly follow the same order as the editing process itself. To further enliven the discussion, throughout the book I refer to various situations and problems encountered during my editing of many films. Several references are made to a TV movie I edited called "Dempsey," about boxing great Jack Dempsey.

After two brief chapters on the history of sound effects and the purpose of sound effects, the book takes the reader through the necessary editing steps: first, spotting the film with the producer (Chapter 3), then preparing effects (Chapter 4), preparing a dialog track for dubbing (Chapter 5), rerecording original dialog (Chapter 6), selecting sound effects (Chapter 7), making use of all available sources of sound (Chapter 8), cutting effects onto film (Chapter 9), building effects into tracks for the dubbing mixer (Chapter 10), doing work on a Foley stage (Chapter 11), and finally dubbing, the last step (Chapter 12). The book also deals with other issues important to the sound effects editor, such as the proper organization of an effects library (Chapter 13), responsibilities of the editor's assistant (Chapter 14), and how to prepare films for foreign distribution (Chapter 15). Chapter 16 shows how the experienced sound effects editor can pick out the effects of a particular film, such as Clint Eastwood's *Dirty Harry*, and analyze its overall effectiveness.

· 1 ·

A Brief History of
Sound Effects

The history of sound effects editing is interesting. In the early days of sound dialog was the only concern. Some films from the 1930s had complete holes in the sound track. People opened doors, walked about, and handled a newspaper, among other things, in silence. Slowly, as the art and technical skill of filmmaking developed, the desire to excel and the drive for perfection impacted upon sound effects editing.

THE FOLEY STAGE

As in all other forms of modern life, technology has had a profound influence on sound editing. A prime example of this is the advent of the Foley stage. Named after the sound editor who first devised the system, a Foley stage (detailed in Chapter 11) is a recording room where the film is projected on a screen, and footsteps along with other sound effects are fabricated to go with the action.

Perhaps the most prominent example of Foley stage at work is the "Mission Impossible" TV show. Anyone who ever saw an episode will remember the detailed work of setting a bomb, opening a safe, or arranging some mechanical device with all its accompanying clicks, ratchet noise, and other sounds. Those sounds were all from the Foley stage. Certainly, they could have been cut from a library; however, the time and money budget would have precluded the required expenditure of man-hours. The Foley stage allowed the editors to do all the effects in approximate sync with each peculiar bit of action in minimum time.

MGM did all the paw sounds for *Lassie* on the Foley stage—a first. One of the major differences between today's films and the old films is where they are shot. In the 1930s and 1940s most films were shot on a stage. The production sound recorder was given time and equipment to properly mike the set. If an actor rose from his chair and walked about the room talking, the stage was miked from one side to the other and one end to the other. In other words, each move, *e.g.*, footsteps, a glass being set down, was recorded live action and in sync. The amount of effects needed to cover the action was minimal because of the beautifully recorded production track. Today most action is shot on location or on practical sets. The actors in many cases are wearing lavaliere mikes, *i.e.*, tiny mikes on a pin or chain around the neck that record the voice but very little else. Therefore, the same scenes, completely recorded in the early days of filmmaking, now require footsteps, the sound of opening the bottle of champagne, pouring the drink, setting the bottle down, *etc.* A great deal of library editing or Foley stage work is needed to cover a small bit of action that would have been totally recorded in the original production sound track under the sound stage conditions of yesteryear.

LOOPING

For most of the early years of sound, the act of looping was unknown. Looping is the replacement of original dialog because of background sound or poor performance. A film shot on a soundstage does not require looping, but the same scene shot on a beach or at another practical location may present a problem. For instance, if a western is being shot on the back lot at Universal Studios, every line is automatically looped for the simple reason that the Universal Studios back lot is on the flight pattern for Burbank Airport a few miles away and also near the Hollywood freeway with its heavy traffic noise. A famous story concerns Kirk Douglas who as *Spartacus* rode over a hill on horseback to confront the legion of Romans. In the background, however, a small piece of the Hollywood freeway could be seen . . . and heard. It was optically eliminated from the release print.

Many of the actors in Universal westerns are day players (actors hired and paid for one day only), and because the studio does not want to call them back and pay them for a day of looping, they are automatically looped anytime they record dialog on an exterior set.

The amount of air traffic today compared to the 1930s and 1940s is drastically higher and more likely to interfere with recorded dialog on exterior scenes.

The art of film was much more controllable under stage conditions than it is today on practical locations. Years ago when a couple had a conversation on a beach the scene was probably shot on stage with a fake ocean or with rear scene projection behind them. Quality dialog resulted. Today that same scene will be shot on a real beach with fluctuating waves and all the other inherent problems very likely necessitating looping.

The art of sound effects editing has become more complex with technologic advances in the same way that the art of auto mechanics has expanded from the needs of a Model T to the requirements of a fuel injected or turbo charged modern automobile.

MAGNETIC TAPE

The advent of magnetic tape has helped a great deal. When working with optical track in the past, if you cut an effect and it was not quite right you could not put it back together. Today working with magnetic tape is much easier. To shorten a crowd applause effect, for example, you can simply splice out a few feet in the middle of the effect so that it begins and ends properly. In the past working with optical track, the same operation was much more complicated, time-consuming, and probably less effective.

PRODUCTIVITY

The amount of time allowed an editor to do the same amount of work has shortened more than proportionally. When I started at MGM the first show I worked on was a comedy movie starring Frank Sinatra called *The Tender Trap*. The entire department went into a screening room and watched the show. The head of the sound effects department then meted out reels to the editors, giving each weeks to complete the work. An editor might spend half a day selecting a door close. Today that editor would probably have one week on the same reel. For television the same reel would be done in no more than three days.

Whereas the mechanics of effects editing have progressed enough to cut in half the required time for any reel, the employers expect (and usually get) the same work done in one-third or one-fourth the time it took in the early years of filmmaking; consequently, productivity has increased beyond the mechanics.

THE INDEPENDENTS

The rise of the independent sound effects company contracting work to the producer has also had an effect on the evolution of sound effects editing. In a constant war for business the various companies strive to satisfy the producer by better and more thorough sound effects. Picking the most minute action, the sound editors strive for perfection.

In the early 1930s footsteps were mostly ignored. As the years passed footsteps within a few feet of the camera became standard. Today some sound editors are reluctant to ignore footsteps a block away from camera. In the 1930s a door close might be ignored, whereas today a cigarette lighter across the room at a busy cocktail party scene might be featured.

Some independent sound companies are spending time to retransfer and replace every bit of production sound. The reasoning is that the original production sound has been run so many times that the emulsion on the magnetic tape track has deteriorated, resulting in diminished sound quality. For a major Dolby sound film there may be a grain of value to this time-consuming practice. For television, however, it is certainly an exercise in futility.

Competition among the independent sound companies is desperate and has resulted in an increased attention to detail. Overall this has resulted in a better quality of sound editing.

· 2 ·

The Purpose of
Sound Effects

The function of sound effects is threefold: (1) to simulate reality, (2) to add or create something off scene that is not really there, and (3) to help the director create a mood.

SIMULATING REALITY

In a western barroom fight our hero is hit over the head with a whiskey bottle. The bottle is, of course, fake. It becomes *real* with the addition of an actual glass bottle crash from the sound editor's library. In gun battles the weapon is actually loaded with blanks and what is called quarter loads, which means one-fourth of the normal amount of gunpowder contained in a real bullet. The actual sound is just slightly louder than a cap pistol until the sound editor has completed work. These are but two of the more obvious examples of the sound effect taking a fake bit of theatrics and making it real by adding a real sound. You see it, you hear it, you must believe it.

CREATING ILLUSION

Creating illusions was one of the biggest additions to the art of film by sound. A man and woman walk into a cafe. Several other people are sitting at various tables deep in conversation. The main couple sits at a table, and a series of close-ups for their conversation are presented. By adding the sound of the off-scene diners the audience is convinced that they are still in the cafe. Obviously, the producer does not want to pay a group of extras to sit off camera. The sound editor places them there with his crowd walla (an industry term for the sound of people talking without hearing specific words).

A woman is sitting in her living room. The door opens and her husband walks into the room. With the addition of a few sound effects, it is possible to inform the audience that he has driven up to the house, parked his car, walked to the door, and used his key to unlock the door. None of this was shot. It was all an illusion created with effects.

A safari makes its way through the jungle. The sound editor cuts a lion roar. Not only has he placed a lion in the film where none exists but he has also placed the safari in danger.

$$1 \text{ Effect} = \text{Illusion} + \text{Mood}$$

I worked on a TV production called "Dempsey" about the former champion boxer. We had a long scene in his dressing room prior to his first championship fight. With effects we created an entire arena of people, yells, cheers, boos, and bell rings for the start and end of the rounds. The entire illusion of a fight taking place off-scene was created.

MOOD

A cowboy sits around a small campfire. The mood of a campfire is warm. Add an off-scene owl and it becomes lonesome. Instead, add a wolf howling in the distance and it perhaps harkens danger. Cut a gunshot and you are telling the audience that another human is nearby. Is he friend or foe? Anticipation, possible danger, fear, joy—all are being evoked. Which should you be feeling? Move forward in your seat because surely the film will soon tell.

A pair of lovers is caught in the midst of an argument. Suddenly, the woman turns, walks to the phone, and lifts the receiver. You can cut the first phone ring just ahead of the reaction and there is nothing unusual. Cut two or three rings before she reacts and you are telling the audience that she was so involved in the argument she did not even react to the phone until the third ring. That is revealing about her mood and depth of involvement.

A leading lady awakens in the morning snuggled in her bed. The sound of a distant train whistle makes it a lonesome scene. Replace the train whistle with the sound of kids playing outside, and the audience perceives an entirely different emotion.

The leading man drives up to a house. As he parks, we hear the sound of a small dog yapping. No particular danger is perceived. Inside is probably a child or an old lady. Change the small dog yapping to the sound of a vicious Doberman, and the mood is again changed.

The sound effect is a major part of the magic of Hollywood. The plywood wall becomes real when the sound of a big wood crash is added. The avalanche of large boulders (actually foam rubber or styrofoam) is real and menacing because of the sound effect.

The purpose of the sound effect is to (1) make the fake real; (2) direct the emotion of the audience in the desired channel; (3) enhance the size of a scene; and (4) create the illusion of that which is not there at all.

To illustrate the uses and purposes of sound effects, consider a specific sequence from an imaginary film, and follow it through the process. The sequence is as follows: Mr. and Mrs. Smith are asleep in their bedroom. Awakened by an alarm clock, they rise, shower, dress, and have breakfast; Mr. Smith then drives off to work.

1. Alarm clock. The director vocally cues the actors to awaken, but the sound editor must cut an alarm clock. Is it an old-fashioned alarm clock or a clock radio? If a clock radio, are they awakened by a buzz or by the radio itself? If by the radio, then we need a voice doing a news or weather broadcast. Then again, the radio could come on with music, in which case the sound editor is not needed and the music would be the responsibility of the music editor.

2. Off scene we should hear a dog bark, kids at play, an occasional car pass by, and perhaps a car horn as a neighbor's ride comes to pick him up. In other words, we use the off scene effects to set the mood. If it is a poor neighborhood, we would have more noise, perhaps the sound of a train passing by. A very wealthy neighborhood would require much less off scene noise.

3. A wood door opens and closes as Mr. Smith enters the bathroom.

4. Open shower door.

5. Turn on shower.

6. Close shower door.

7. Turn on bathroom faucet, and later turn off bathroom faucet to match picture.

8. Open medicine cabinet.

9. Foley movement of medicine cabinet paraphernalia (see Chapter 11).

10. Foley brush teeth (see Chapter 11).
    We now cut to Mrs. Smith in the kitchen.

11. Cut the sound of the shower to a new track for off scene level.

12. Refrigerator door opens and closes.

13. Crack two eggs.

14. Frying sound of eggs.

15. Time the off scene shower turn off so that Mr. Smith has time to dry

off, put on his robe, and come in for breakfast.
Cut to Mr. Smith driving to work.
16. Sound of car engine steady.
17. Sound of specific car passing by.
18. General traffic background.
19. Parking garage background.
20. Mr. Smith's car in and stop.
21. Car door opens and closes.
22. Elevator steady as Mr. Smith rides up to his office.
23. Elevator door opens and closes.
24. Light office background (typing, phones, voices).
25. If Mr. Smith happens to be a doctor arriving at the hospital, we might have a hospital background, including elements such as doctor pages, call bells, and other sounds inherent to a hospital corridor.

A simple scene such as that just described can, and usually does, require a great many sound effects. What has sound effects done for this scene? Well, possibly we never saw the shower but convinced the audience that Mr. Smith was showering with a sound effect. We have established the nature of their neighborhood with our selection of off scene sounds. Does Mr. Smith work in a small office or in a large one? Again, we have answered that question for the audience in our selection of sound effects. Perhaps Mr. Smith arrived in his office with no visible clues to the nature of his occupation. The sound effects will inform the audience with the use of a steel mill background, a lumber mill background, or any sound effect specific enough for the audience to identify.

The sound effects have clarified some things about Mr. and Mr. Smith including their finances, his job, and what they ate for breakfast. They have disclosed the man's occupation without the producer having to shoot the actual location. The sound effects have taken away the stage feeling of the bedroom or office and made the audience feel that they are mid-city. The careful use of sound effects has set up the audience to accept what we want them to accept. The sound effects have saved the producer the cost of shooting the shower, the kids playing in the street, and the steel mill, which can amount to several thousands of dollars in time and film.

SILENCE

There are times that a scene is played without the sound effects. A montage generally does not have any effects. The music carries the story. Some directors prefer to leave out some effects while stressing one specific effect, thus using the sounds available to create the mood of the scene.

A scene opens with two cowboys sitting by a campfire. The fire is actually a rather dull and annoying sound. Often the scene is dubbed as follows: (1) play the fire up to a normal level; (2) as we pan, or cut into a close-up, we fade the fire down until it is almost if not totally gone. The sound of the campfire was used to set the mood, and then it got out of the way of the dialog.

I once did a show in which a bomb threat caused the evacuation of an airline terminal. Once empty the camera panned about the terminal without a human to be seen. It occurred to me that a telephone ringing incessantly would stress the absence of people. We mixed the scene with the phone loud and reverberating about the terminal. It was spooky and effective.

Two lovers are involved in a serious conversation. Suddenly, the woman turns, goes to the telephone, and lifts the receiver. The editor may choose to cut the phone bell to the actress' reaction or have two or three bells before her reaction, thus informing the audience that she was so engrossed with the conversation she ignored the phone for a while. The audience learns something about the characters and their state of mind. The sound effect has been a theatrical tool used by filmmakers to direct the manner in which the audience perceives the scene and the actors in it. The sound effect is not only a mechanical tool (*e.g.*, fire a gun, cut a gun shot), it can also be employed creatively and artistically. It is the singular ability of the human mind to be creative, and that creativity is what makes a good sound editor better than another sound editor.

· 3 ·

Spotting

A WORD OF CAUTION

Before beginning a discussion of the editing process, it is important to recognize that there is a lack of standardization in the film business. From studio to studio basic principles change. For that matter, many things change from editor to editor. Some put their start mark on the 0 frame of the synchronizer and others put it on the #1 frame. The system and methodology change depending on the studio. When a reference is made to a *studio*, that well might mean independent sound editing company as well as a major film studio as we have come to think of them.

On one job the specific sound effects may be selected by a supervising sound editor and handed to an assistant for editing. On others the selection of specific effects may be left to the individual editor. On one job the supervising editor may spot the reel and explain exactly what effect to cut and where. On others the sound effects editor may be handed a reel of film and given no specific instructions.

In the case of some major films a departmentalization of effort exists. For instance, one editor supervises the ADR, another supervises the Foley, while a third concentrates on effects. In the case of some major films the amount of work becomes too much for one editor to supervise; thus, the responsibility is divided among two or more editors so that each area can get the proper detailed consideration.

The extent of the inconsistencies in the film business can be confusing. For example, some studios operate with the sound side of the magnetic tape up and others with it down. Almost all use what is commonly referred to as the academy start mark, which appears 12 feet before the first frame of picture. Universal uses a start mark 20 feet ahead of the first frame of action.

The wide range of discrepancies makes it difficult to make too many absolute statements. Still, the important part of this text remains the same regardless of the particular methodology of the employer. Caring about work, creativity, thoroughness, and accuracy are the hallmarks of an adept sound effects editor. Two things (at least) are constant throughout Hollywood: film runs 90 feet per minute and good work is good work.

SPOTTING A MOVIE OF THE WEEK

Major shows, such as a *Star Wars*, are careers unto themselves. Very few people ever have the opportunity to work on such a large-scale project. To use such a film as an example would require a much larger book. Thus, for our purposes we will concentrate on something with a greater chance of being encountered, a TV movie of the week. The point of reference will be that of the supervising sound editor. You may be working either as an independent contractor or on payroll at a major studio. In either case the work needed on the film and the basic approach to that film would be the same. The job has been contracted, the financial arrangements made, and it is time to *spot* the show. *Spotting* is the process in which the supervising sound editor, in concert with the producer and/or director, screens the show to discuss what is both needed and wanted by all parties involved. The desired result of this conference is for the show to arrive at the dubbing stage with everything required to create the final sound track exactly as the producer envisions.

There are several methods used to screen the film at this point. In addition to the projection room described at length here, there is the flatbed editing system and the video system.

It is not uncommon to spot on a flatbed editing system. This system, actually quite functional, allows for noting exact footages and is fast. Spotting also occurs on video system, but this can be undesirable for lack of getting exact footage notes. The mechanical system on which you spot does not affect the nature and purpose of the session. The easiest and most functional system is still a projection room. The projection room should have the capacity to run the film forward and in reverse. It should also have a footage counter so that the editors can take specific notes.

It is during this spotting session that the producer first relies upon the expertise of the editor for advice regarding which dialog can be saved in the dubbing and which dialog should be looped. The producer often asks for effects or combinations of effects that the editor knows will not work to the end desired by the producer. At this point it is the editor's role to advise the producer on how to accomplish this end with a different effect or combination thereof. For example, if we see an actor stub his bare toe on a table leg and we cut a small sound to cover the action, it will *work*. However, to tell the audience with that same sound that the actor we do not see has stubbed his toe off camera will not *work*. The effect is not specific enough.

During spotting it is the supervising editor's responsibility to get as much information as possible from the producer. Important questions include where we are in any given scene in relation to other activity. A restaurant scene might warrant questions regarding how busy the crowd should be. An active walla track may be desired to convince the audience that there are more people in the restaurant than we see. Conversely, the scene might be very quiet with just the hint of walla from the other diners.

OFF CAMERA EFFECTS

The editor gathers from the producer his or her ideas regarding the off camera effects. For example, a scene takes place with two actors in a serious discussion in their living room late at night. Does the producer want to hear crickets from outside? An occasional car passing? A distant siren? A dog barking down the block? Without getting the producer's requirements for this scene, we run the risk of arriving at the dubbing session without enough off camera effects. In contrast, too much time can be spent adding some of the aforementioned effects when a producer dismisses the effects as unwanted noise during a serious late night scene. It is important to keep in mind that film is a subjective art and the producer's subjective views are paramount.

An example of not spotting a show properly took place years ago when I was working on the live action "Spiderman" TV series. One episode had a scene in which Spiderman grabbed a dirt motorcycle and chased the villains into a film studio back lot. It was apparent the motorcycle he was riding was a dirt bike. We cut the sound of a dirt bike, not having specifically discussed what the producer wanted. In the dubbing room the producer suddenly announced he wanted the motorcycle to sound like a Harley-Davidson. To satisfy this demand would have required a full day on the editing bench changing the tracks, not to mention the extra costs in dubbing time. The audience was composed of young children, many if not most of whom knew the sound of a dirt cycle and knew that it sounds vastly different from a Harley-Davidson; they would have been amused. A compromise was reached, and we continued dubbing. As a compromise, the dubbing mixer equalized the sound effect we cut, taking out all the high frequencies and adding as many low frequencies as we could.

The importance of getting into solid agreement with the producer before starting to actually edit the film cannot be overstressed. For instance, in a car chase, does the producer want the cars to sound real, or does the producer want to go for dramatic value? The car chase in *Bullitt* was one of the best ever done. Nobody objected that the engines used were very high revving engines not at all realistic for the cars we saw on screen. Undoubtedly, the editors of *Bullitt* and their producer were in agreement on what the cars were going to sound like before they started cutting.

A number of years ago I worked on a film for the Levy-Gardner-Laven Production Company. The film was *White Lightning* and starred Burt Reynolds. Arthur Gardner paid for a car with a modified engine; an entire day was spent recording on a major southern California racetrack. Before we started cutting Mr. Gardner came into the cutting room and listened to the sound of the car. Working for a producer like that was a pleasure. The end result is a better sound track at less cost for the producer.

In contrast, I recently worked on a science fiction horror film. The producer would not come into the editing rooms to listen to any effects beforehand. Keep in mind that nothing could be more subjective than the sound of something that is not real to begin with. We cut the film using effects we thought worked. Unfortunately, the result was a total disaster. The first time the producer heard the sound we selected for the spaceship door opening, he hated it. We had to change approximately 60 spaceship doors. The producer ended up hating most of our sound effects. It was not that we were wrong or right; it was simply that the producer had a perception of the sounds he wanted but never conveyed it to us. The result of that lack of communication was a long, expensive dubbing session that was annoying for all involved.

During the spotting session the subject of looping will arise. Looping is the process of rerecording the production dialog on a stage (see Chapter 6). Nothing should be looped that can be saved by proper editing techniques and/or dubbing room electronics. The producer often turns to the editor for advice on looping. As a norm, looping results in a loss of performance. It is a difficult process for the actor, expensive for the producer, and time-consuming for the editors. It is important for the editor to have the expertise to properly advise the producer. If the editor suggests that a particular line of dialog will be okay when in fact it will not clean up decently on the dubbing stage, the recriminations will fly in the direction of the editor. Again, it is vital for the editor to be in agreement and total understanding with the producer about the show and all the effects going into it *before* leaving the spotting session.

· 4 ·

Preparation

The show has been properly spotted and now you are back in the cutting room. Every editor has his or her own way of working. There are no rights or wrongs as long as the end product is good. The next step consists of running the show through the moviola and making exact and specific notes (Figure 4.1). Place a reel of picture and a reel of track in the moviola at the start mark. Zero out the footage counter on the moviola. Running picture and track together, you spot to specific footage and sometimes frames and specify exactly what effects are needed. Earphones are needed because the speaker on a moviola offers poor reproduction and one can never reasonably judge sound quality or hear low-level intrusions in the sound track. Two passes should be made through each reel. On the first pass you may spot the problems in the dialog track and order production reprints to correct them. During this first pass you may also want to take notes on looping. During your second pass you can concentrate on the effects needed and spot Foley. It is virtually impossible to do a comprehensive job of spotting a reel when trying to think of all the aforementioned elements at once. For the dubbing session to be successful and for the producer to be satisfied with the work, thoroughness must be maintained during the passes. If you have any doubts about an effect, cut it. It is much easier to close a pot and leave an effect out of a mix than it is to run out to the change room during the dub and cut an effect that should have been cut in the editing room. If you have any doubts about an effect, cut

Figure 4.1 A working sound editor at his bench. Moviola in foreground. Picture and three sound tracks being wound through synchronizer. Note the often used sound effects lined on rack above bench: a few basic doors, crickets, and a couple of basic clicks. If you were doing a gun battle film, you might fill rack with shots, riccos, and gun clicks.

an alternate. Does a particular crowd sound too active? If you are not certain, cut an alternate, less active crowd. It shows the producer that you care. This approach can of course be taken to an extreme at which point you are cutting the show twice.

JUDGMENT

An important element in successful sound editing is judgment. The judgment involves knowing what is needed and what is not. An editor with poor judgment who cuts twice as many effects as needed is ineffective in that more time, material, and dubbing stage time are used at a loss to the production company. The most glaring example of overkill I can recall occurred on a show we did titled "Gettysburg," a short subject shot at the site of the famous Civil War battle. There were no moving objects on screen. The camera moved about the landscape as a voice-over told of the battle in detail. One editor cut 55 tracks to cover one reel. It could have been properly covered with half that number, but the editor's lack of judgment and self-confidence led to cutting way too much. In today's fast-paced, high-productivity, post-production film world, overkill is too costly.

Often the budget of a show will determine how detailed the effects will be. A major film with a $20 million budget will require a more detailed and exacting sound effects job than an independent film shot on a $1 million budget. Also, the former film will have a much larger sound effects budget than the latter. The budget will go a long way in determining how detailed the spotting must be. A sample TV movie of the week with a moderate level of detail (allowed by the budget) is examined. The movie is "Dempsey," the life story of the famous heavyweight boxing champion of the world. It was done first as a four-hour TV movie and later cut down to a foreign theatrical release. The length of that show would be too cumbersome for our purpose; consequently, I have condensed it into one reel of film. This sample reel will run the gamut of the entire show; it consists of eight scenes as follows:

12 Feet

1. Young Jack is getting boxing lessons from his older brother in the barn on their farm.

187 Feet

2. Jack is helping his mother dry the dinner dishes as he tells her he is going off to seek fame as a boxer.

273 Feet

3. A montage as Jack roams from barroom to barroom earning $5 and $10 for bare-knuckle fights.

305 Feet

4. Jack and his brother discuss Jack's future as they sit eating lunch in the rock quarry where the older brother works.

354 Feet

5. Jack's first real fight in a small building with approximately 100 spectators.

415 Feet

6. In what serves as a dressing room, Jack receives a telegram from a New York City boxing manager offering to get him big fights if and when Jack comes to New York. (Off scene another match is in progress.)

THE CUTTERS
3849 STONE CANYON AVENUE
SHERMAN OAKS, CALIFORNIA 91403

FEATURE-_____DEMPSEY_____REEL#----1----------

Footage	Description	Effect#	Box#
12	Birds	Bird 296	
12	off scene barnyard bkg.	Bkg. #1241	
14-	Punch---cover all punches thru 187feet	Fight #3,4,5,819	& 823
18-	Body fall on dirt	Fight #22	
73	Horse whinney	Animal #99	
172	Boxing glove fall onto dirt	Fight #904	
187	off scene crickets	Animal 316	
198	set dish on table	Glass 645	
219	same		
265	start montage (no effects)-		
273	bare knuckle fight	Fight #33 & #34	
273	Men yell for fight	Crowd #310	
279	Body Fall onto wood	Fight 486	
279	Men cheer victory	Crowd 248	
288	cont. Men yell for fight	Crowd 310	
292	Bare knuckle fight	Fight #33 & #34'	
299	Body fall onto table	crash 811	
299	Cheer victory	Crowd 248	
305	off scene mine bkg.	Bkg. 1945	
305	start horse passby	Hoof 6674	
305	sweeten with pic & shovel track	Work 310	
328	sweeten 1912 car passby slow	Model A #6	
336	horse passby	Hoof 6674	
354	100 people yell for fight to 370	Crowd 115	
355	cover all punches to :370	Fight 33,34,41	
370	old time fight bell	Bell 68	
388	same		
388-	start crowd yelling for fight	Crowd 115	
:370	crowd idle between rounds to 390	Crowd 411	

Figure 4.2 Spotting notes. Some version of this form is a prime example of well-prepared spotting notes. The left column indicates the footage at which the effect appears. The center section describes the effect needed. That is followed by the exact

THE CUTTERS
3849 STONE CANYON AVENUE
SHERMAN OAKS, CALIFORNIA 91403

FEATURE- _____ DEMPSEY _____ REEL#__1_ Pg.#2___

Footage	Description	Effect#	Box#
396	Crowd Cheer knockout	Crowd 415	
415	off scene crowd yell for next fight	Crowd 116	
416-	cheap wood door close	Door 226	
423	off scene fight bell		
423	off scene crowd idle between rounds	Crowd 411	
511	big railroad station bkg.	Bkg. # 727	
514	Station announcements	A.D.R.	
552	off scene train whistle	Train 1242	
577	Start traffic for off scene	Traffic # 108	
586	cut traffic to on scene track		
586	Old car pass at 30mph	Packard #7	
597	2nd. old time car by	Winton #3	
603	start horse trot by pavement	Hoof 3886	
603	start carriage sound passing	Wagon #53	
640	off scene big crowd yell for fight	Crowd #436	
662	old fight bell off scene	Bell 68	
662	Crowd idle between rounds	Crowd #89	
687	cut crowd idle to on scene track		
723	Old fight bell	Bell #68	
723	Crowd yell for fight	Crowd #311	
729	Do all punches	Fight 711,712	
748	Add cheer for knockdown	Crowd 383	
750	add track wilder cheering	Crowd 119	
788	five or six bells rapidly after 10 count	bell #68	
749	Body Fall on mat	Fight #780	
748	Big grunt	Voice 611	

effect number recommended by the supervising editior. Some version of this form is important as a tool to organize a reel and as a communication tool between the supervising editor and crew.

511 Feet

7. Jack arrives in New York's Grand Central Station and is met by the manager.

687 Feet

8. Jack's first big time fight in a New York arena with several thousand spectators.

Figure 4.2 demonstrates the cutting notes on this reel; the form used is one I designed years ago. The first column shows the footage of the effect. The second column describes what effect is needed, and the third lists specifically, by number, what effect from our library is to be used. The notes were compiled as the reel was run through the moviola.

KITS

The next step is for our assistant editor to make a *kit*. A kit consists of all the material needed to cut the reel. The editor then need only pick up a box and he has all the effects needed. There are many places (both independent studios and major studios) where the editor must select material and pull the prints from the library.

Among the material needed are the production reprints, copies of the original sound track needed to fill ambience, and otherwise prepare dialog tracks for dubbing. Production reprints are discussed in Chapter 8. Other needed materials are the effects, a copy of the looping notes, and the Foley notes.

· 5 ·

Dialog Tracks

The dialog track is the sound that was recorded during the actual filming. It is often referred to as the production track. When the film editor constructs the picture, the production sound track (dialog track) is edited to match the picture.

It is a good idea to work with dialog tracks before moving on to sound effects. Once the necessary time and effort have been spent on the dialog track a familiarity with the show will be attained as well as knowledge of what is needed to complete the sound track.

One of the most difficult responsibilities of a sound editor is preparing a dialog track for dubbing. This is a complicated procedure because each problem requires its own considered solution. The dialog track is the copy of the original sound that the film editor cuts along with the picture. It comes to the sound editor in various states of disarray. There are directors' voices, miscellaneous stage noises, differences in the recorded level of dialog within a scene, clipped words of dialog, and other problems.

It is the responsibility of the supervising editor, while spotting the reels, to note every problem and order a production reprint to help the individual editor solve the problem.

During the filming of a show the negative comes out of the camera and goes to the laboratory for development into a color work picture, which is what the film editor cuts. The original ¼ inch copy of the production sound goes to the sound company and is transferred onto 35 mm magnetic tape. It is the assistant film editor's responsibility to put picture and track into sync and code them. Coding is the process of running both picture and track through a machine that puts a tiny series of letters and numbers along the edge of the film at 1-foot intervals. Those numbers allow the film editor to keep picture and track in sync while cutting.

To get the material needed to fix a problem with the dialog track, the sound editor gets the code number off the edge of the track, which indicates which of the ¼ inch tapes the original sound was recorded on. A new copy of that entire scene is made for the sound editor. That copy is referred to as a production reprint.

A variety of problems can be encountered during the working of a dialog track. Some possible problems with the solutions are numbered below.

1.   A director's voice. There is no such thing as silence. Every scene has some ambience. It may be very low as in a living room set or very loud as in a scene shot on a busy downtown street corner. The track must never go dead, *i.e.*, have no ambience. A director's voice is in the middle of a dialog scene. The voice must be removed, but the ambience of the room must be replaced. Take a production reprint of the scene and go to the spot in that scene closest to the spot of the director's voice. Get a piece of ambience from that reprint with no other voice or particular movement and replace the exact length of the director's voice with *clean* ambience.

2.   Two people are arguing. The film editor has had to cut from one actor to another before the first was finished talking. In the production track the last word of the first actor's line is clipped, although not completely. Get a production reprint of the clipped word, complete that word on the master dialog track, build a second dialog track (usually called B dialog), and move the second actor's dialog to the new track. The result will be overlapping dialog, which is true to life for that scene.

3.   There is a scene with four actors in one room. Actors A and B are on one side of the room laughing. In the middle of their laughter the film editor has cut to actors C and D who are in a serious conversation. The laughter of A and B cannot just cut off with the picture cut. They are still relatively close and within earshot of the camera. Get a production reprint of actors A and B, continue their laughter on a new track identified as dialog B (so that the dubbing mixer can drop the level of their laughter on the picture cut because they have gone somewhat off scene), and build a third track identified as dialog C onto which you will cut the conversation of actors C and D. The simple scene is now on three dialog tracks rather than one, but it is properly cut so that the dubbing mixer can control the levels of each element.

4. An exterior scene involving two actors is shot on a street corner. The master two-shot is done in the morning during heavy traffic hours. The close-ups are done at midday when the traffic flow is much lighter. Every time the film editor has cut from master to close-up, or *vice versa*, there is a drastic change in the background level of the traffic. First, split the dialog, *i.e.*, the master track stays on the dialog A track, and build a dialog B track for the close-ups. Then order a production reprint of the master angle. From that enough footage of the heavy traffic flow is available which does not have dialog in it to make a loop; the longer the loop, the better. From that loop you can transfer enough footage to continue the heavy traffic flow through the entire scene on the A track. In some instances you will allow the low traffic level in the close-up angle to come and go with the dialog. The problem with *filling* is that the low level traffic may come and go. Generally, it is not necessary to fill the B track. Also, it is recommended to have a traffic effect running in an effects unit. The dialog mixer and the sound effects mixer can then collaborate on covering any bumps in the traffic background that may occur by using the sound effect traffic as a leveling agent.

5. Two actors are yelling at each other in a jailhouse scene. For some unknown reason one actor has a great deal of reverberation (natural echo) in his track and the other actor has none. When the film editor makes a quick cut from the reverbed actor to the other, there is an audible cut-off of the reverb tail in the track. To solve the problem, the two actors are split onto two separate dialog tracks. Then reprints are ordered of every line with incomplete reverb. When a specific word is incomplete in the production track, the same word should be found in the reprint and the entire clipped word replaced, allowing the reverb to have its natural tail-out in the A dialog track while the second actor is starting to talk in the B dialog track.

6. A scene is filmed in a two-shot master with two actors talking as an automobile drives by very close to the camera. The film editor cuts from the master just as the auto is passing the mike. The close-up has no auto. In the production reprint the continuation of the auto passing is covered with actors' lines that are not in sync with the way the picture is cut. There is no solution; however looping can be easily performed by simply throwing away the dialog and rerecording it on the loop stage. There are times when looping is not possible. For instance, the actor is out of the country busy on another film, or perhaps the producer does not wish to pay the additional costs to bring the actor in for one line. It is then up to the sound editor to make the scene work as shot. First, split the scene with the actor whose line is covered by the car left on the A dialog track and the second actor's line with no car on the B dialog track. Then wipe the track with the car. Re-

member that no sound ends instantly. Every sound has some natural trail-off. *Wiping* is the process of creating an artificially quick end to an otherwise immediate ending sound (Figure 5.1). Given the aforementioned scenario the best approach would be to take the piece of track where the auto ends and lay it on the editing table with a piece of paper tape angled across the sound stripe, creating a severe angle of magnetic tape from full width to nothing in the space of three or four frames. Once the tape has been properly positioned across the sound stripe, a dissolving solution should be applied to a cloth to wipe the magnetic tape strip. This will probably sound better in the moviola than an instantly ending automobile although not terrific. The next step is to cut an automobile passing by into an effect track with the heart of the passing by happening at the same time as the abruptly ending auto in the dialog track. The sound effects mixer will then use that automobile to help cover the problem in the dialog track by playing the effect automobile slightly louder than the automobile in the dialog track.

Whenever a piece of track from a production reprint is spliced into

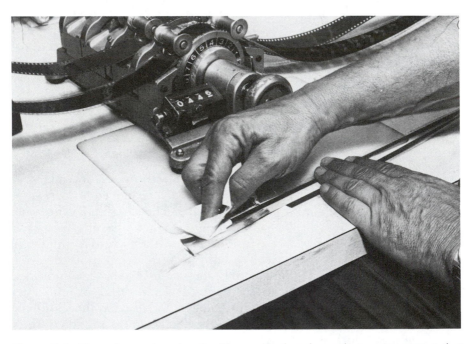

Figure 5.1 To make a wipe, lay the film on the bench, apply tape to protect that portion of the mag stripe you wish to maintain, and (using a solvent) eliminate that portion of the sound stripe necessary to create a natural end to that sound.

the original sound track, it is important to make the splice at a point of sound impact. In problem 5 (above) the entire word was replaced, starting with the impact of the voice. In cases of a slight difference in the level of the original and reprint, the ear will not hear it as readily when the change in level occurs on the impact of the word or other action. An actor walks out of a room, closes the door, and then there are a few feet of air before the next actor talks; that few feet of air may have a director's voice or some other unwanted sound. The solution is to fill the unwanted noise from the point of the door close. If splicing is attempted air to air, *i.e.*, room ambience to room ambience, there will often be some minor difference. The ear will never hear the splice if the fill is started with the door close. Then filling should continue to the impact of the incoming voice.

The proper preparation of dialog tracks is just one aspect of a sound editor's duties. Changes in voice quality within a scene are noticeable and annoying. The quality of a voice within a scene might be likened to skin tone colors. They must remain consistent.

Problems with the sync can be adjusted right on the stage. An effect can be equalized to give it some slightly different sound. A loop can be hung to change an effect or to blend with what has been cut. Dialog, however, must be in sync and the tracks built so that the mixer can handle it properly.

FUTZ

In a scene with a girl talking on the phone to her boyfriend, the proper way to prepare the track is to have the girl on an A track, the boy split onto a B track, and all the dialog coming over the phone on a C track so that the mixer can *futz* it. Futz means to give a normally recorded track the quality of a telephone, radio, or other electronic apparatus.

Over the years I have encountered many sound editors whose only measurement of the work is whether or not dubbing was successful. If it was not rejected by the producer from the dubbing stage for a missing effect, poor choice of effect, or an out of sync effect, then it was acceptable. This attitude is unprofessional. Often a dubbing stage is working under a tight schedule. The producer may well accept a reel of sloppy sound effects because time is not available to call the editor for the changes needed to make it a good reel. The fact that the reel was not rejected does not mean it was good.

A skillfully prepared dialog track is now ready to dub. Background changes, perspective differences, futzing, and level changes have been split. Yet one element of the dialog not accounted for remains—automated dialog replacement, or looping. This element is addressed in Chapter 6.

IMPORTANCE OF SAVING PRODUCTION SOUND

One of the first big films I worked on was *Guns for San Sebastian*. I was handed a reel of picture and track and a set of spotting notes by the supervising editor. One scene involved a group of Federales riding their horses into a large Mexican cathedral. The scene lasted 40 feet. There was one good piece of production sound (of about 10 feet) with the sound of the horses' hoofs on the tile floor. It was a great sound made more so because it was real. The spotting notes from the supervising editor stressed that he wanted me to make the 10 feet of production sound work for the entire 40-foot scene. That bit of production sound was much better than anything we could cut from the library or make on the Foley stage. Thus, good production sound is always better than anything that can be made.

Another good practice is to move that good production sound from the dialog track to a sound effects unit, thus saving that sound for the foreign version. The sound will go into the sound effects channel of the master dub. Left in the dialog track, it will go into the dialog channel and be lost for the foreign version.

· 6 ·

Automated Dialog Replacement (Looping)

Looping used to refer to the act of rerecording original production dialog for any of several reasons, including poorly recorded dialog, the producer's desire to change a performance, or the need to add and/or change a word.

Poorly recorded dialog was usually the result of unwarranted background noise making its way into the sound track, *e.g.*, a western filmed outdoors with the sound of an airplane flying overhead. Occasionally, a producer does not like the performance of an actor in a given scene. By looping that actor and saving the other dialog, you can change one performance in a scene without altering the others.

The term looping came from the physical act of removing the sound track containing the particular line of dialog and splicing the two ends together. The track then formed a loop, which would go around and around on the dummy as the actor kept giving new readings of the line until you had a *take*. A take was achieved when the editor was happy with the sync and the producer happy with the performance.

A new, modern, and more efficient system was developed, called automated dialog replacement (ADR). The system incorporates a computer, a projector, and a sound dummy. The editor notes the exact footage the line to be looped starts and ends. Those footages are fed into a computer, which then operates the system, which in turn places all three elements on interlock, runs the film and track down to the line to be looped, opens up for recording at the proper time, and closes at the end of the line. There

Figure 6.1 An ADR stage. The film is projected on the screen. The actor stands at the podium with typed script of lines to be rerecorded. The ADR or sound editor sits at the table. The actor and editor each wear an earphone allowing them to hear the original sound track. A series of three beeps occurring 3, 2, and 1 foot ahead of the line cue the actor. The editor hears the original line over the earphone and the live actor through the uncovered ear. It is the editor's job to hear sync.

is a series of three beeps occurring 3, 2, and 1 foot before the start of the line. Those beeps give the actor a rhythm to start reading.

The actor stands in front of a podium wearing a set of earphones into which is fed the original line. He or she watches himself or herself on the screen and speaks into a microphone, attempting to duplicate the original line of dialog (Figure 6.1).

PROGRAMMING

The first step in the ADR process is programming, *i.e.*, to note the start and stop footage and frame of every line to be looped. It is important to be sure the lines are written exactly as they were spoken. A script is often of little use because actors often read lines slightly differently than they were written. A slight, meaningless change here or there or difference between the typed script and the words actually spoken will create problems. Indicate grunts, stammers, and any other utterances besides just words (Figure 6.2).

COMPANY: __Your Company__

PRODUCTION: __Your Film__

REEL NO.: __1__ Sheet No __1 of 1__

Microphones: _____

Equalization: _____

Start Time: _____ Finish Time: _____

Mixer: _____

#	CHARACTER	DIALOGUE	FOOTAGE Start Stop	CHANNELS 1	2	3	4	MIXER NOTES
1	Joe	Sam, where were you last night?	START 110.5 STOP 113.6					
2	Sam	I wastired so I stayed home	START 113.11 STOP 116.6					
3	Joe	You missed quite a party	START 116.10 STOP 118.2					
4	Joe	Helen got loaded and put on a real show	START 118.12 STOP 121.1					
5	Sam	She does that every party...keeps it live	START 121.14 STOP 124.0					
6	Betty	Last night she went really nuts	START 124.9 STOP 126.15					
7	Betty	I think she has a real drinking problem	START 127.6 STOP 129.1					
8	Sam	She just likes the attention..no class	START 129.2 STOP 131.12					
9	Joe	Attention is one thing but she is getting out of hand.	START 132.8 STOP 135.4					
10	Betty	It's been getting worse since her divorce	START 135.8 STOP 138.2					
11	Helen	You have to make him pay my alimony on time.	START 363.5 STOP 366.2					
12	Lawyer	He has been served all the papers the court allows.	START 366.15 STOP 369.12					
13	Lawyer	The next step is to threaten jail	START 370.0 STOP 372.1					
14	Helen	That's fine with me..serves him right.	START 372.1 STOP 374.10					
15	Helen	If he pays me late, I can't pay my rent on time.	START 375.2 STOP 377.14					
16	Lawyer	You are spending money rather freely Helen.	START 378.6 STOP 380.9					
17	Secretary	(W.L.) Mr Richmond on the line for you.	START 381.0 STOP 383.1					
18	Lawyer	Helen, give me a few days to work this out...call me Monday.	START 384.5 STOP 387.13					
19			START STOP					

Figure 6.2 An ADR cue sheet. As prepared by the ADR editor, the sheet gives the character's name, the exact dialog to be looped, and the footage for start and stop. The recording mixer keeps notes on the printed takes in the columns to the right of the page. Each column can contain a reading of the looped line. The editor must know which reading the producer wants in the show. It is imperative that we indicate exactly what was said and not what might be in the script. All pauses, grunts, uhs, and stammers must be indicated. The producer has the option of making changes; your job is to show exactly what was said.

A major problem is long sentences. The longer the line, the harder it is to sync. Long lines are often read without any break. For example, "Good mornin Mr Smith howareya doin today I hope all is well." This line is delivered with no breath and no pause, thus difficult for the actor to sync. The idea then is to find the longest pause, however short, and note the end footage and the opening footage of the continuation. Program it as two lines. Then mark the exact frame of picture where the second half of the line starts. Using a long ruler or other straight edge, draw a line diagonally across the picture. As the film is projected that line will move across the screen. When it hits the edge of the screen it will act as a cue to the actor. If the actor decides to read the line in one piece, the cue will help him hit the second half in sync. If he decides to do the line in two pieces, the beeps will serve as a cue. If the line has been programmed as two lines, the system can be set to open and close first for the open half of the line and then for the second half. If the line is to be done as one, the middle numbers can be ignored and the opening of the first line and the close of the second line can be used. The actor's job is difficult because he read the line originally on the set, working with other actors, and was in mood. Now you want him to recreate the dramatic value of the line on a stage alone while concerning himself with beeps, sync, and the lack of ambience. Anything that can be done to help the actor helps the film. The goal is to make it as easy as possible to read the lines while retaining their dramatic value.

RECORDING ADR

The main function is to check sync. If the line is recorded out of sync, it will be hard or impossible to cut it into sync. You should be able to tell if the actor is reading the line too fast or too slow. Your help with speed should be appreciated. If there is a slight pause in the line that the actor is not quite catching, let him know. If he is stressing a different word within a line, let him know. The different reading can actually affect the meaning of the line. Often by stressing a word he did not stress in the original he is causing himself to get out of sync and cannot ever get back in sync unless he stops stressing that one word.

Under normal circumstances you will have a take when the producer is happy with the dramatic reading of a line and you are happy with the sync. Without a producer or director, the responsibility falls entirely on your shoulders. The best you can do under these circumstances is to get exactly the same reading on the loop line as the actor gave on the original. Occasionally, the actor will say, "I want to lower the emotion of this line a bit," or "I want to stress a different word than I did on the original." You have no real authority over the actor. Your best approach here is to record the line as the actor wants. Then tell him you should have a duplicate reading just in case. You can later ask the director which reading he wants.

When looping a line with a change, try to get the key words, those beginning with the letters T, B, D, F, K, P, and V to hit sync. It is harder to see exact sync with words beginning with letters such as S. The hard impact words are more definite and easier to see when out of sync. When recording a line with a change from production, let the producer know that it might well look slightly out of sync on the dubbing stage.

On the stage you will be recording onto a 35-mm piece of full coat mag, which means that you have four separate channels of track to work with. Suppose that the actor does a reading that is not bad but not great. You can save that line (in case it is the best he ever does) and tell the recorder to go to channel two. Record the line again and then decide which of the two you want. Neither? Go to channel three and try again. You can actually have four different readings of the same line before you either decide to print one of them, or if you want to record again, you erase one of the four you are holding.

As the editor on stage you have several functions. While sitting at a table with an earphone over one ear, you listen to the original line of dialog and must make the judgment on sync. If the line is out of sync, you must ask the actor to try again. This can be difficult, *e.g.*, when the actor has read the line 10 or 15 times and it is still out of sync.

The ability to keep things loose and the actor in a good mood will help everybody. It can help the actor get a good reading, it helps you get a reading in sync, which will cut easy, and it speeds up the process.

WILD LINES

You may get a list of wild lines the producer wishes to record on the ADR stage. A wild line is one either off camera or coming from an actor with his face turned away from the camera, *i.e.*, a line with no specific sync. You should program all wild lines to the footage they will appear in the show. You will often be asked to record them together and with no relation to their appearance in the film. The problem then is to keep a record of where they must be cut. If you record them where they are to appear, they are easier to keep track of. Also, by recording them in sequence you can hear them against any lines that appear just before or just after, and you can make a judgment on their quality. Do they work against the lines around them? If you have a scene in which all the actors are speaking very loudly or with great emotion, any wild line recorded to go into that scene should match level or emotion with the production dialog.

EDITING ADR

One method widely used to edit the loops into dead sync is to run the picture with the original dialog track and mark impacts. For instance, "Good morning, Mr. Smith." The first impact word is the G in good. That is the only real impact in the line, but you could also mark the M of Mr. and the M of morning. You can then run the loop line and edit it so that the same impact words fall on the same frame of picture that they did on the production track.

If you have a scene with ten lines of dialog, run the entire scene and mark all impact letters. This is not automatic; you must still run the loops and check that they look right. Some lines will be easy. You can just slide it a few sprockets and they fall in. Others, which were not recorded in good sync, require some work. Again, consider the line, "Good morning, Mr. Smith". You match the G at the head of the line, but running to the Smith you find it looks a frame or two late. Take a frame of silent track out between Mr. and Smith. Run the line first and make sure it does not sound odd. Usually, one frame will not harm the rhythm of a line. By eliminating that one frame you cause the Smith to come that much sooner and it will now look in sync.

The reverse approach would be taken if the Smith looked a frame early. Splice a frame of leader into the track between Mr. and Smith causing Smith to come later. If the line was recorded way out of sync, you may have to add or subtract more than one frame between Mr. and Smith. An alternate approach would be to add a frame before Mr. and another before Smith.

Occasionally, you encounter a line that cannot be fixed as described. For example, "How you doing Joe?" The line is read as one long word with absolutely no pause and no place to add or eliminate as much as a sprocket. The actor read the line faster on the loop stage than he did on camera; you compromise. When editing you find that starting how in sync causes the Joe to come two frames early. The compromise is to start how a frame late and allow the Joe to come one frame early. One frame out of dead sync on most dialog will pass; it is called soft sync. Two frames out of dead sync are usually noticeable. If you must be out of sync, always be out early rather than late; it is not as apparent.

When cutting into a line of dialog be careful. If you are eliminating a frame, save it until you have run the line and made sure that the elimination works. You may find the line sounds strange and want to replace the frame. You must never cut into sound. Any time you add or subtract a frame from within a line it must be in silence. Cutting into a word will create a noticeable glitch.

When cutting loops double and triple check yourself. Run the looped scene two or three times and check every line carefully. You do not want to go to dubbing with a line out of sync. Going to the change room during dubbing to add an effect is bad enough, but taking a loop line to the change room in the midst of a dubbing session is torture. It is time-consuming, tedious, and embarrassing. Working in an unfamiliar cutting room, pressed for time with people asking how long you are going to be can often cause you to make a mistake.

ACTORS

Over the years I have worked with many actors and actresses, and you will no doubt encounter many types of actors. Some of my favorite stories relate to the loop stage and the actors I have worked with. One of the most amazing was Ken Curtiss who played the Festus character on the old "Gunsmoke" TV series. Festus was a character who invented most of his dialog as he went. His delivery was irregular, and he stuttered and stammered continually. Ken could take a long line full of irregular pauses and stammers, listen to the original track two or three times, and then give a dead sync reading on the loop stage. Fred Astair was also unique. Most actors listen to a line once or twice and are ready to try recording. It then takes an average of three or four attempts to get a take. Fred Astair would listen to every line five times. When he said he was ready, he was ready; one take did it—dead sync. Supposedly, it had something to do with his dancer's timing.

Many actors are afraid of the loop stage. They do not like having to recreate, under sterile circumstances, a performance they gave working with other actors on a live stage. It is a difficult task. Add to that the problem of sync. They have to concentrate on the beeps and try to get the line exactly as they did on stage. Then they have a producer, director, and an editor all telling them what they are doing wrong.

Once the dialog tracks have been prepared and the required ADR recorded, it is time to get on with the portion of your work that is the most obvious—the sound effect.

· 7 ·

Selecting the Effects

In many cases you will cut a reel from a kit supplied by the assistant editors as selected by the supervising editor. Under those circumstances, do not hesitate to go to the supervising editor when you run across an effect you think does not work.

This chapter deals with the problem of selecting your own effects. To do this you must know the library and have good taste and judgment.

The only requirement for an effect is that it *works*. You will hear that term often. It implies only that when the effect is viewed against picture, it appears to be real. Basically, when viewing a film and hearing the sound, you believe what you see. Usually, a sound effect must be really bad or really wrong not to work.

SATISFACTION WITH THE EFFECTS

As a sound editor of great taste and judgment, you must satisfy yourself in the cutting room. If the effects you cut are right for you, they should be right for the producer, and you will feel confident if asked to defend them. It is best to start with an effect that according to the description in the library index seems like what you are looking for. For example, if you are doing a scene with Mrs. Smith driving the kids to school in the family station wagon, you will not want to use a Model T, a Porsche, or even a

Volkswagon. You will look for a generic American car with eight cylinders of reasonably recent vintage. Likewise, if she is driving a Volkswagon, you will not use a 1983 Oldsmobile; the effect must be reasonably close.

If you are doing a gun battle you will not want to use a rifle for a pistol or a rifle shot recorded in a canyon with lots of reverb for a shot in a living room.

CROWDS

Crowds are the most difficult effect to cut. Crowds have age, size, emotion, ethnic character, location, and many other individual characteristics. For example, when in need of applause, does the crowd number ten people or 500? Are they applauding politely or with great enthusiasm? Is the crowd outside at a ballpark or in an auditorium? Each of these and many other questions must be answered before you can select the right crowd applause effect. The walla in a high school cafeteria will sound very different than in a five-star restaurant.

Some terms of classification in most libraries are as follows: walla, moderate talking; babble, more active than walla; murmur, less active than walla.

HORSES

Horses' hooves are another difficult effect to cut properly. If you look into any major sound effects library, you will find horse hoofs classified under single horse, two or three horses, and posse. There is no difference between five horses and ten horses. Then the pace of the horse comes into consideration. Walk, trot, lope, canter, and gallop are the general pace classification. Then you get into dirt, cement, and wood as surfaces upon which one, two or three, or a posse can be walking, cantering, trotting, loping, or galloping.

It is important to get the right effect. If you have a dog barking on camera, you must use a dog bark from the library that sounds like the dog you see, *i.e.*, a dog of approximately the same size; you would not use a poodle for a great dane. The dog must also be of the same temper. There is a difference between a casual barking and a vicious barking.

Another example involves church bells. Is it a church on the corner in your neighborhood, Big Ben in London, or an adobe church in some small Mexican village? Each is different; each used incorrectly will jar the audience and make them aware they are watching a film and not a slice of life. It will work against the film. You would not use the ticking of an old, cheap alarm clock for the grandfather clock in some mansion.

BACKGROUNDS

Backgrounds are a very important tool. Our library has hundreds of background tracks catalogued as interior or exterior. Backgrounds are such effects as police station, bowling alley, gambling hall, prison mess hall, harbor, airport, western street, desert, jungle, construction site, *etc.* They are great tools because a good background fills a scene and requires only to cut specific or close-up effects.

There may be a scene with two people talking along the East River in New York City. The production track has only the two voices. We see very little activity on the river behind the actors. A good harbor background will fill the scene. It provides a water quality, a few boat horns off in the distance, and probably a boat engine passing. It fills the scene and informs the audience of the locale. It also lends size and scope to the scene.

Many low-budget pictures are saved with good effects and especially good background tracks. A cheap set with a scientist in his white smock and a few vials of bubbling water becomes bigger, more real, and believable with the addition of a good scientist's lab background track. For example, the hero, a race car driver, pulls his car into the pits. All we have to this scene is two or three men, a car, and a properly dressed race pit set. What does the sound editor do for the scene? He adds a race track background to convince the audience the race is continuing off camera. He adds a crowd yelling track to convince the audience the 2000 extras are still watching the race. He adds a race track pit area background track to make the audience think there are men working on other cars just off camera. Three background tracks have taken the place of ten race cars, 2000 extras, a few mechanics, and a few cars.

Much of your work on any given show is in the dialog track. Much may be in cutting loops, which when done well are not apparent. Only the gunshot, the crowd applauding, or the car chase lets the producer know what kind of job you have done. Those effects should be selected carefully.

THE FAKE

One aspect of sound effects editing that is challenging and fun is faking. When you have exactly the effect you need for a piece of action, it becomes a mechanical operation to get that effect in sync with the picture and build it into a track. What happens when you do not have the perfect effect? You either get it from another library, make it, or fake it. Faking is the most fun. For example, I once had a scene with a bicycle riding past camera in the distance. I did not have a bicycle passby in the library at that time, so I used an old-fashioned press. It had the right rotary quality and the rattling, metallic sound we needed. If the effect had been in close-up, I might have been in trouble, but played low in long shot it worked.

Faking is particularly challenging when faced with a problem on the dubbing stage. There was a scene in which an actor turned on a kitchen tap and poured a glass of water. This action occurred in extreme close-up and in our black and white dupe was not visible. The point of the scene was someone entering the room 30 feet away. The producer wanted to know where the sound of the kitchen tap was. A backup of about 40 film boxes filled with various loops for just this type of problem was available, and a loop under the water category was located. The dubbing mixer played it low; too much level would not have worked. The process took 3 or 4 minutes and the sound was okay. As it turns out, the material used was the bow wash of a navy destroyer. It would have been preferable to have seen the kitchen tap and cut it properly in the editing room.

Hospital rooms, even intensive care units with all the scopes, *etc.*, do not actually make noise except in a movie. A beeper or some other type of effect in a hospital room is usually needed, however. Almost anything will work as long as it is nondescriptive and technical sounding. Faking provides an opportunity to be imaginative and creative.

It is a rewarding occasion when you have an opportunity to work with a producer who is concerned enough with sound effects to help you get the right ones. For example; a few years ago I worked on a film called *Outlaw Blues*. It starred Peter Fonda and Susan Saint James. The film's climax was an outboard motorboat chase along a southern river delta. There was not a boat in any library that had the size and menace we needed to sell this scene. The producer arranged with a major manufacturer to bring one of their race boats to a local lake and we spent the day recording; they worked great. We now have the best outboard motorboat library in Hollywood . . . and have used it to trade with other libraries for material we needed from them.

Film school students often get the experience of recording sound effects because most schools do not have comprehensive libraries. It is an excellent ability to possess.

Faking, like anything else, has its limitations. My favorite story about faking was a film I worked on at MGM years ago called *Kansas City Bomber*. It starred Raquel Welch as a roller derby star. A point in the story was one character who played the nasty skater. He was born and raised on a pig farm and when the fans wanted to get on his case they would chant "suey," as in hog calling. Luckily, the company had to shoot some additional scenes that called for an arena full of people. On our request the time was taken to have them chant "suey" and record it for use in the film; it worked great.

Because sound effects are subjective, you must ultimately use your own judgment. If you are in doubt about a given effect, cut an alternative. Give the producer a choice of how to play an effect. It shows your interest in the show.

MULTIPLE EFFECTS

Many times a single effect will not cover the action properly. Cut two or three effects, which when played together will work. For example, you have a big fight scene where some people are cheering and some are booing. This effect is difficult to find in a library. The obvious solution is to cut one cheering crowd and one booing crowd. Let the dubbing mixer put them together, in proper balance to each other, and you will have a terrific crowd.

As another example, suppose the producer wants a car to start and pull away from the camera sounding old and exhausted. If such a car exists in the library, great, but if there isn't one, start with a basic, old car and sweeten it. Sweetening means adding to the basic effect to enlarge or otherwise alter its sound. You may add backfires on a separate track so that they can be played at an appropriate level during the mix. You may get creative and add something as outlandish as an old washing machine (as I did once in an Elvis Presley film). I had an old printing press and all kinds of junk built into the tracks. We ran through the scene five or six times during dubbing, and each time the director laughed, he enjoyed it, and it was funny.

As a young editor I was looking through a library to find a small flag flapping in the wind. Someone suggested I look under boats and find a sail flapping. I did and it worked.

For the sound of lifting one of those long wooden arms you see at the old European border points I once used the sound of Cheetah the monkey hand-cranking a disinfector. It sounded wooden, it had a ratchet quality, and it worked. No matter what it was, it worked played against the picture. Once again, a fertile imagination is an invaluable asset in selecting effects.

· **8** ·

Ordering Material
from the Library

Learning to order material from the sound effects library is an important skill for an assistant sound editor. If done properly, the assistant can ensure that the editor has all the material and in sufficient quantities to complete the work. Of lesser importance, but not to be ignored, is the assistant's impact on the production budget. Ordering material and transfer time are expensive. Running about wildly ordering material well in excess of actual requirements can add greatly to the cost of the project. Putting aside the few big-budget shows on which it seems nobody cares about cost, the vast majority of shows are cost-conscious.

Every studio and post-production company will have its own system for ordering material. It is therefore impossible to address every eventuality you may encounter. As an insight to this part of the process, I will deal with the problem within our system.

The assistant takes the notes on each reel and checks the cutting library to see what we have. The cutting library is a copy of the master library created when we overtransfer. For example, I may need one door close #860. When the assistant transfers that effect he will not transfer just one close but 20 or 30. Once the master has been pulled and is threaded on the transfer machine, it costs almost nothing to transfer a few extra. When I have taken the one I need from that roll, the balance is filed in the cutting library. The next time I need a door close #860, it is there.

When our assistant finds we do not have a copy of door close #860 in the cutting library, he must pull the master copy and make a transfer for his editor. If door #860 is one we use very often, he may choose to transfer 400 to 500 feet of the material so it will be on hand for a while. If door #860 is some unusual, infrequently used door, he may choose to transfer only 50 feet.

Backgrounds require some thought. Sounds such as traffic, surf, birds, a railroad station, and office backgrounds must be transferred to fit need. If the scene for which the effect is needed is 300 feet long, we must transfer at least 310 feet; 210 feet will not do much good. It will cause a delay and lose time. If the background is a commonly used one, we may want to transfer 600 feet—300 to be used on the show at hand and 300 feet to be put into the cutting library for the next time it is needed. Once you have spent the time to pull the mast from the library and threaded it onto the transfer machine, it costs little to transfer that extra 300 feet, certainly much less than to go through the entire routine again next week.

Transfers are costly but much less so than time. A good rule is to transfer more rather than less. The editor should know how much to transfer.

The reprint system for Universal Studios called for making two copies of every reprint the editor needed. On one show I had a 1-foot director's voice that had to be filled. The supervising editor noted the scene number required to make the fill. Nobody noticed that the scene was 800 feet long. As a result, I received 1,600 feet of transfer for a 1-foot fill.

Stock effect (those used almost continually) will often be held in inventory. For instance, bird #296 is the general, backyard bird background we always use. We often roll in 300 to 400 feet of this material in a scene. The cheapest way to have that material is to transfer 1,000 feet at a time. The assistant pulls the master (one advantage of a loop master) and transfers 1,000 feet at a time. Whenever we get down to the last 200 feet, we automatically transfer another 1,000 feet. An editor can just grab a reel of bird #296 and roll in all that is needed.

We take the same approach to several, often used effects such as a basic wood door open and close, general traffic, surf, wind, and a few others. When transferred from a loop the material can be rolled into a reel without listening. If, however, we maintained a 100-foot print master, we would have to make ten transfers, which takes five times longer, and then splice several prints together to cover a 400-foot scene. Then each splice must be heard to ensure there is no bounce in levels from one to the other, a time-consuming process.

ADDITIONAL SOURCES OF SOUND

In addition to the sound effects library, the editor may have two additional sources of sound to choose from: production sound and sound you record for the film.

The first source is production sound, that sound recorded with the film during actual filming. This consists primarily of dialog. On occasion the production recorder (the person recording sound of the stage or set) will have the opportunity to record a wild track. For example, a scene is being shot in the midst of a working print shop. When the production recorder wants to record the sound of that print shop without extraneous noise, *i.e.*, director's voices or other sounds of the crew, that wild track will eventually be available and of great value. It is the actual sound of that actual print shop and will be better than the sound of a print shop coming from the library.

Many wild tracks will be of locations, machinery, or movement that cannot be obtained from a library. It makes for a better sound track.

A second source of sound is that which you or a member of your staff record directly for your film. As mentioned elsewhere in this text, I once did a film (*Outlaw Blues*) that ended with a motorboat chase. I could not find the sound of a motorboat in any library in Hollywood that sounded right. We were able to get a major manufacturer of outboard motors to bring a boat to a local lake, and we spent the day recording the boat for use in the picture.

You may want to take quality recording equipment out and record some sounds for film. Maybe you want the actual sound of traffic on Sunset Boulevard; you can simply go out and record it. The sound of a crowd leaving the opera after a great performance and the background around Los Angeles International Airport are sound effects that can be recorded for your own library.

Recording to order is an excellent method of keeping your library up to date. If you record traffic on Sunset Boulevard, you are not going to put it into one film. You are going to add that effect to your library first and then use it in the picture. The same holds true of wild tracks. Although recording to order is a valuable source of material, you will still have to rely on the sound effects library for the majority of your material.

· **9** ·

Cutting

Cutting refers to the physical act of syncing a door close to the action on screen, placing a start mark on a 1000 foot roll of blank film, rolling the picture and blank film (called leader) through a synchronizer to the spot of the action, and splicing the door close into the leader. Later the film will be placed in a projector, the leader on a sound dummy, and interlocked so that when the action of the door close appears on screen the sound of the close will come from the dummy. A typical editing bench is shown in Figure 9.1.

Hopefully, you have been able to spend enough time looking over the shoulders of other editors to know how to cut.

The show has been thoroughly spotted, the kits prepared, and you are sitting at the moviola. Now what?

I remember putting my first reel of picture in the moviola and running down to the first effect noted by the supervising editor. It was a truck starting and pulling away from the camera. I took the proper material from my kit, got it into sync with the picture, and stopped. It occurred to me that if I put a number on the effect just ahead of the first sound and the same number on the corresponding frame of picture, I could later splice the mag into a reel of fill leader. Every effect is numbered with the matching number on the frame of picture opposite to which the effect is to be cut. Roughly 20 years and hundreds of reels later I still cut the same way. Most editors work the footage system. Once you have an effect in sync with the

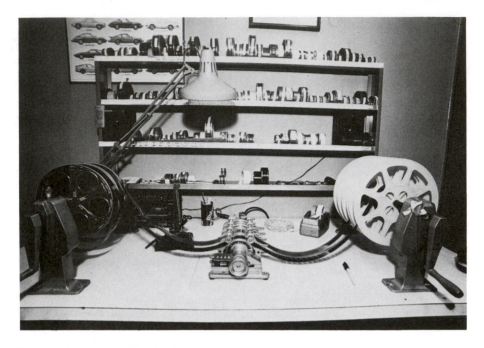

Figure 9.1 A typical editing bench.

picture, you note the footage on the moviola counter and write that number on the effects. Later you place a reel of film leader in a synchronizer, run down to the appropriate footage, and splice the effect into the leader.

Over the years I have had many editors observe my system. The footage system allows for several ways of getting out of sync. You can read the footage incorrectly, you can write it on the effect incorrectly, and when you are cutting it into the leader you may again read the number on the mag or the number on the synchronizer incorrectly. Any of these will result in your being out of sync on the dubbing stage, which can be most embarrassing. I highly recommend my system. In this text we will work with the numbering system.

NON-SYNC EFFECTS

Throughout the rest of the chapter, I will explain how to cut various effects from the spotting notes used for the TV movie "Dempsey." Please refer to Figure 3.1 for the order, type, and number of effects. The first effect asked for is birds. Birds are one of many effects I classify as a non-sync effect. A door close, gunshot, or car passby must match the action of the picture. They are sync effects. Birds along with such effects as traffic, general backgrounds, wind, and rain obviously have no sync. It is therefore not necessary to cut those effects in a moviola. I mark the frame of picture where I want the birds to start and the frame where they end. I can later roll them into a reel of leader in the synchronizer. At 12 feet I mark "birds" on the proper frame of picture and move on.

SYNC EFFECTS

The first sync effect we have is a punch at 14 feet. We have many punches running all the way (as noted) to 187 feet. To do this scene requires both a mechanical understanding and some creative judgment. You do not want every punch to sound the same. The supervising editor has selected five different punches to be used in the scene. I would start with punch #819, get into sync with the first punch, and mark the #2 on picture and track. I would make a small mark on the picture telling me I have covered that punch. I would splice leader onto the end of the punch and run down to the third or fourth punch and again cut an #819 and mark the picture. Continue through the scene covering approximately one-fifth of the number of punches with #819, marking the picture and splicing leader in between the punches. At 187 feet you stop, roll up the material you just cut, and hang it in your trim bin. Now go back to the beginning of the scene and with punch #823 repeat the process. Avoid creating any pattern to the punches. Occasionally, you might use the same effect twice in a row and then not again for ten punches. Use the various punches in a random sequence. When you are finished with the scene you will have five rolls in the trim bin numbered 2, 3, 4, 5, and 6. Each will go into a separate track of leader. On the dubbing stage the result will be a fight of almost 2 minutes in length with a lot of punches, but they will sound different and random and the scene will work.

You have finished the tedious process of cutting the punches. Back at 18 feet, stop the picture on the exact frame where the body hits the dirt. Take effect #fight 22, run it on the sound head, and stop at the first impact of sound. Put the sound and picture heads on interlock and look at the body fall. Does it look right? When you are satisfied that the body fall on dirt is in sync and sounds right, number it and the picture with a #7, cut it from the roll, and hang it in the trim bin.

At 73 feet the supervising editor has called for an off scene horse whinny. You must assume he spotted the effect where it will not interfere with dialog. Follow the same procedure with the horse whinny as with the body fall on dirt except for the number, which is now #8.

At 172 feet you need a boxing glove fall to dirt—a rather unique effect that may not be in your library. If the effect asked for, fight #904, does not work to your satisfaction, see that the effect is added to the Foley notes and have it made there.

At 187 feet you will note the end of the birds and barnyard background, which both began at 12 feet and the start of the crickets. Note that the only difference between on scene and off scene crickets is the level at which they are played in the dub. The off scene will be noted on your dubbing cue sheet as a clue to the dubbing mixer.

Continue working your way through the reel as explained above. At 273 feet you start men yelling for fight. That is a continuous yelling of appropriate size and intensity. When you start the cheer at 279 feet the yell should continue for a few feet so that the dubbing mixer can fade it out under the cheering. At 288 feet while the cheering should end with the visual of the picture, the yell should begin a few feet before the cheer ends so that a sound segue can be made. There should never be a hole, *i.e.*, a point at which there is no crowd sound. Nor should it seem like the cheering ends and the yelling begins instantly. In real life there would be an overlap, some men starting to cheer while others are still yelling and *vice versa*. The conversion from cheer to yell and back again should be smooth, necessitating overlapping the effects.

At 305 feet there is a mine background. In looking at the scene we find there is also a horse passby, an old car passby, and then another horse spotted. If the mine background is busy enough, the scene may be well covered with those effects. However, you naturally desire to put your own mark on your work. Although you always do everything asked for in the spotting notes, you are not limited to those effects alone. If you have a fine idea or if you see something the supervising editor missed, that is fine. You should not be restricted to the notes alone. If you think the mine scene would be helped with a horn for a change of shifts, cut it. Any time you think a given scene might be helped with an effect the supervising editor did not think of, cut it. Put your own creativity to work. It is one of the most rewarding parts of editing.

To cut a horse passing camera you run the picture to where the horse is closest to camera and stop the picture. Then run the sound effect to where the sound of the hoofs is loudest. Put picture and sound heads on interlock, back up a few feet, and look at the effect. If it feels right, it probably is. If you are looking for an exact sync point, be sure to place the sound of a hoof landing at the point where the first on camera hoof lands. The same principle applies for a car passby. Place the effect where it is loudest at the point the car is closest to camera. Juggle it until it looks and feels right.

At 354 feet we have our first actual ring fight. The scene starts in action, so we start with a crowd yelling for the fight. Cut the first fight bell where spotted, run the yelling crowd about 5 feet past the bell, and start the crowd idle for between rounds right after the bell. In the dubbing room the mixer can make a segue from yelling to idle, making it sound like one crowd quieting down between rounds. Reverse the procedure for the start of the next round. Start the yelling with the bell running the idle 5 feet longer for another segue back from idle to yell.

At 410 feet we cut to the dressing room of the arena. We want to create the illusion of another fight going on in the ring outside the dressing room. Use the same crowds but on different tracks so that they can be played lower, possibly with some reverb to create an off screen flavor.

REVIEW

Sync effects must be cut in the moviola and numbered so that they can be built into tracks later. Background effects such as birds, traffic, waves, or restaurant voices, *i.e.*, non-sync effects, can be marked for exact start and stop points on the picture and rolled into a track when building. In the case of a picture dissolve from a downtown street corner to the beach, we must overlap the traffic into the beach scene and start the waves in the street scene so that the dubbing mixer can make the sound dissolve match the picture. When possible time off scene effects such as a horse whinny, kids playing outside a kitchen, a car horn, or whatever so that they do not fall at the same time as dialog or other on scene effects.

At 577 feet we start a track of off scene traffic as the two men walk through the train station and approach the door. The dubbing mixer will fade into the traffic slowly as the men near the door. At 586 feet we cut to the street as the two men exit the station. At this point we end the off scene traffic and continue it on a different (on scene) track so that the level (volume) of the traffic can jump up on the cut.

At 586 feet and again at 597 feet we have older cars passing by. These are to be approached in the same manner as the horse passby at 336 feet, *i.e.*, run the picture until the car is at its closest point to the camera. Then run the sound track to where the car is closest to the mike. Put the moviola on interlock, back up to the beginning of sound, and run the picture and track together. If the doppler effect of the car passing matches the point where the engine of the car passes the camera, you have succeeded.

The doppler effect is a term we should deal with. The sound of a passing train or airplane changes between the approach and once it has passed. That is a doppler effect. One of the most common is the train crossing bell. It has a very different quality as we come upon the crossing and another after we are pulling away. The point of change in the bell is at the moment of closest proximity. The basic approach to a passing effect is the same whether that effect is a car, train, plane, horse, barking dog, or even a woman's footsteps.

Notice at 687 feet we have a crowd idle track going from off scene to on. The material you will use is the same. To make it work, *i.e.*, for the level of the crowd to jump up on the picture cut, you only have to cut the effects on two different tracks. The mixer plays the off scene at one level and the on scene slightly higher. When the picture cut happens the crowd level will jump up.

At 748 feet we want to add a crowd for a knockdown. The best approach to this effect would be to continue the crowd yelling, which started at 729, and add the cheer on another track. This gives the dubbing mixer the latitude of playing the cheer right over the yell or lowering the yell to make room for the cheer.

It is important to cut and build a reel so that the dubbing mixer can deal with it in the easiest possible manner. Building a reel is discussed in Chapter 10.

It is a good policy to check items off the spotting notes as they are cut to avoid missing anything. Thoroughness is important even though the spotting notes may be shoddy and the supervising editor's judgment incorrect. This is the best time to write the reel, thus possibly avoiding a situation in which the producer is kept waiting on the dubbing stage.

Use your judgment and add your own creative touch to make the reel live. You should have a picture in mind of what the reel will sound like when the dubbing mixer is finished. You cut element by element, but you must be able to hear it all together in your mind.

TRICKS OF THE TRADE

There are many tricks of the trade, too many to list here, but some are listed below.

1.   When you cut an effect start as close to the first modulation (sound) as possible. Some effects have a low level hiss on the track. If you cut three or four frames of apparently blank mag ahead of a gunshot you might well hear hiss–bang, hiss–bang on the dubbing stage.

2.   When cutting multiple impact effects such as gunshots or thunderclaps, place them on one track with two sprocket holes of blank fill leader in between. For instance, there is a gun battle scene involving a hero who fires off five shots in 3 feet. How do you cut them? Not on separate tracks. All his shots go on one track. Suppose the first shot is not over when the second occurs. The impact of the second shot will be louder than the tail of the first. To hear a clean impact of the second shot, cut off the first, insert two sprockets of fill leader, and cut the second shot. You get clean, sharp impacts with this system. You can cut 100 shots on the same track with this simple method, providing they are all the same shots. If you have one character with a pistol and one with a rifle, each weapon must be on

its own track. The same rule applies to a series of thunderclaps close together. In our sample reel at 788 feet we call for five or six bells to end the fight. The same rule applies. All six bells on one track will be separated by two sprockets of blank.

3.   If you must make a choice, always cut an effect early rather than late. An effect out of sync early is less obvious than one out of sync late.

4.   Chin socks are difficult. Often the actor jerks his head back when the attacker has barely started swinging. Do you match the head jerk or the point at which the fist is closest to the chin? There is no pat answer. Cut it where it looks best; sometimes there is no such place, but do your best.

5.   It is extremely shoddy work when an actor picks up a telephone and it continues to ring for a second. How do you get the ring to stop with a lifting of the receiver? Not with a scissor. The bell has a natural ring off and must not be cut before the natural ending. Two solutions are offered: (1) start the ring where indicated and either pull the rings slightly closer together or conversely space them slightly farther apart so that the last ring before the lift up ends properly; or (2) cut into the last ring and make it shorter than normal so that it ends properly. Either way, have the last ring end on time and always with a natural tail off of the bell.

6.   Remember that all sounds have a natural tail. A sound can start instantly such as a gunshot, but it never stops instantly. The only time you can cut off a sound is on a picture cut with a change of location. A car passby tails off into the distance; it must fade away, it never just stops. If you are standing at the beach and listening to the waves, the sound of the individual wave always fades out rather than just stops. Each sound is like a wave and must have its natural fade.

7.   In spotting an off camera effect avoid laying over dialog. When a reel is spotted and an off camera effect is noted, it will be in between dialog.

8.   A car comes to a stop. Sync the ending, the point at which the car actually stops, and back the effect in.

9.   There is a difference of opinion as to cutting such things as a car passby to a car steady. I cut on the picture cut, whereas others like to overlap the picture cut two or three frames with the passby to avoid a sharp change in sound. Some editors overlap backgrounds a frame each side of a picture cut. For example, we cut from the beach to a downtown intersection. Some editors start the traffic a frame or two early and run the surf a frame or two into the downtown scene. I prefer to cut on the frame. It is a matter of subjective opinion.

10.   Remember that there is a difference between a car skidding on dirt and one on cement; this difference is often disregarded.

11. Another distraction and often committed error involves the actor who picks up a telephone and has a conversation in which we do not hear the voice of the calling party. That caller then hangs up at his end and we hear the dial tone. If we can hear the dial tone, then why did we not hear the voice? On a recent TV show an actress telephoned her ex-husband. We heard nothing as she announced to the audience that she had his answering machine. Then at the proper time we heard the click of the answer phone hang up. If we could hear the hang up, why did we not hear the voice announcement? Creative license (from at least the producer) is viable, but there should be some attention paid to reality.

12. Do not hesitate to build as many tracks as may reasonably be needed. For example, you are on a landing with a helicopter idling. You cut back and forth from alongside the copter to the hangar 50 yards away. Two-track the copter idle so that the mixer can play the close-up idle louder than the distant idle. To give another example, in a scene one actor's voice has some natural reverb while the other voice is flat. Two-track the voices so that the dubbing mixer can add some reverb to the flat voice and match them. Supposing you Foley a scene with an actor walking away from the camera. After 20 feet we cut to a close-up of the same actor still walking. Two-track the steps so that the dubbing mixer can create a fade to the exiting footsteps and then jump back up to close-up level on the picture cut.

There are times you will encounter an impossible situation. I once had a scene with a car driving up to a hotel during a heavy rain. There was a visible clap of lightning just as the car came to a stop. When I cut the thunderclap where it belonged, it looked like the car was crashing into the hotel. There is no clear-cut solution.

Consider a girl target shooting in the desert. The special effects man set off the ricco three frames before the girl fired the gun. To rectify the situation, cheat each effect a little. Place the gunshot one or even two frames early if it does not look too bad and the ricco two frames late. Consider two policemen shooting at a fleeing killer. Each cop fires twice. Each cop's shots fall in the exact same frame as the other's. The solution is again to cheat a little. Place one cop's shot a frame or frame and a half early and the other's equally late. This is not enough out of sync to be terribly noticeable and far enough apart to sound like two shots. Lastly, consider two men fighting on a bus traveling 25 miles per hour through the streets of Hong Kong. We cut away momentarily and on the cut back to the bus it is at a dead stop. The only logical way the bus could have stopped that fast was to have hit a brick wall. To solve this dilemma, cut a track for the bus steady until we see the bus at a stop. Then also cut a track of a bus to a fast stop. Let the mixer and the producer figure it out on the dubbing stage. Perhaps there will be a lot of heavy chase music and you will hear very little of the bus.

CONCLUSION

After you have carefully and creatively selected your effects and cut them in sync with your picture, you now have a trim bin full of sound effects. (A trim bin is a metal box on rollers with hooks from which you may hang film awaiting further use.) Because you cannot take a trim bin to dubbing, it becomes necessary to build those effects into tracks so that the dubbing mixer can deal with them properly.

· 10 ·

Building

As in the cutting of sound effects, many different approaches to building are utilized.

I will teach my method of building, which goes along with my method of cutting and while not the only method, I think it is the most efficient.

When I have completed cutting my sync effects and marked picture for the roll-in effects I can usually guess how many tracks it will take to build my reel. Let us assume that you are not sure; take two reels of leader (blank film into which we will splice our effects), label them with the proper reel number, and call them Fx #1 and Fx #2. Allow approximately 6 feet of leader, then place a start mark on each reel. Put the picture in your synchronizer with the academy start mark and your two effects tracks with their start marks on the zero frame of 000 footage. Start rolling down, picture and tracks in sync, and when you get to 12 feet you start. Assume you are going to put your birds and barnyard background in tracks 1 and 2, respectively. Take your print of each, run in the moviola to be sure you are in modulation (that portion of a magnetic print that actually has sound on the stripe), and splice the birds in Fx #1 and the barnyard in Fx #2. By checking the cutting notes you know the two effects run all the way down to 187 feet. When you have rolled the effects in check for the exact frame (as you marked it in the synchronizer) where they end, cut them off and splice the leader back onto the ends of the mag.

As film runs 90 feet a minute you want to allow at least 20 feet and

THE CUTTERS
3849 STONE CANYON AVENUE
SHERMAN OAKS, CALIFORNIA 91403

FEATURE _____ DEMPSEY _____ REEL# 1 pg. #1

Start		End	Description	Effect
12	1	187	Birds	296
12	2	187	Barnyard Background	1241
14	3	168	Punches	819
14	4	169	Punches	823
15	5	166	Punches	3 & 4
18	6		Body Fall Dirt	
73	6		o/s whinney	
172	6		glove fall	
187	5	273	o/s crickets	316
198	4		set down dish	
219	4		set down dish	
273	2	278	Punches	fight 33
273	3	278	Punches	fight 34
273	1	283	Men yell .	Crwod 310
279	4		Body Fall	
279	5	286	Cheer	Crowd 248
288	1	291	Men Yell	
292	2	299	Punches	Fight 33
292	3	299	Punches	Fight 34
299	6		Body Fall	
299	5		Cheer	Crowd 248
305	7	354	Mine Background	Bkg. 1945
305	4	320	Horse by	
305	1	354	O/S Digginh	Work #310
328.	2	340	Car by	
336	4	351	Horse by	
354	5	375	Crowd Yell	
355	3	396	Punches	
355	2	396	Punches	

Figure 10.1 Cutting notes. This figure shows a set of well-organized cutting notes. When completed during reel building, the notes may be used to make out actual dubbing cue sheets. During the dubbing session these notes are invaluable. You can follow the action on screen, making certain every effect is played.

THE CUTTERS
3549 STONE CANYON AVENUE
SHERMAN OAKS, CALIFORNIA 91403

FEATURE _____ DEMPSEY _____ REEL# _2 pg. #2_

Start		End	Description	Effect
370	6	393	Crowd idle.	crowd 411
370	7		Bell	Bell 68
388	7		Bell	
388	5	410	Crowd yell	
396	1	415	add cheering	crowd 415
415	4	428	o/s crowd yell	
416	6		door open / close	
423	7		o/s bell	
423	2	511	o/s crowd idle	
511	3	586	Station Bkg.	727
514	1	540	Station P.A.	
552	2	586	o/s train whistle	1242
577	4	586	o/s traffic	
586	5	640	Traffic	
586	I	609	car by	
597	6	611	2nd. car by	
603	7	621	Horse by	
603	1	621	carriage by	
640	3	663	o/s crowd yell	
662	7		o/s bell	
662	6	687	o/s crowd idle	
687	5	728	c.u. crowd idle	
723	7		Bell	
723	4	755	crowd yell	
729	1	660	Punches	
729	2		Punches	
729	3		Punches	
748	6		add cheer	
750	5		more cheers	
788	7		Bells	

preferably more before you cut another effect into either Fx #1 or Fx #2. Obviously, the dubbing mixer will have the level of each of these two tracks set so the birds and the barnyard backgrounds are at a proper volume. If you cut another effect into either of these tracks 3 feet later, he would have only 2 seconds to react and move his pot control from one level to the proper level of the second effect, which is not enough time. When we have cut off our opening effects at 187 feet and want to start the off scene crickets we cannot build them into Fx #1 or Fx #2. The crickets will be built into Fx #5 when we build that track. At 273 feet we have some punches. I would let those go into the same track as the opening punches at 14 feet. Then cut our men yell for fight at 273 feet. I generally try to keep impact effects in tracks with other impacts and long running (background type effects) in their tracks. Thus, we will cut men yell for fight into Fx #1 and at 279 feet cheer victory into Fx #5.

The purpose of the number on picture and track is simple. As you roll your picture and tracks through the synchronizer you will see a number on the picture. You simply splice effect #2 into a track directly opposite the number 2 on the picture. If you were to place all seven tracks for a given reel of picture into a synchronizer and roll down, you would be able to check sync on every effect by seeing that the number of the effect falls opposite the same number marked on the picture. For a roll-in effect you can check by the mark you made on the picture and also be seeing that the effect starts on the proper picture cut. Traffic would start on the cut from an office to the street.

I have always followed the practice of making a scratch dubbing cue sheet after building my reel so I can "see" the entire reel laid out on paper. If there are any effects too close to others, or if I can see where I might combine seven tracks into six comfortably, or any other problem with the way I organized the reel, I can make the change before taking the time to make a final clean and neat cue sheet for the mixer.

We continue running through the reel, cutting as many effects into #1 and #2 as we can, leaving the mixer ample time to reset pots between effects. When we have completed building #1 and #2, rewind, mark Fx #3 and Fx #4, put them in the synchronizer and start again. This time I would cut punches #819 and #823 into the tracks. Then at 198 feet I would cut the dish set downs, at 273 feet the bare knuckle punches, and so on.

On the third pass through the reel, while building Fx #5 and Fx #6 cut those effects not built into the previous tracks.

On the cutting notes form we use the last column marked box to tell the assistant editor where the master of an effect for purpose of transfer can be found. For this exercise I have left this column to mark in which track I would cut each effect. Study the sample dubbing cue sheet (Figure 10.1), and see how the reel can best be laid out. Similar effects should be done in the same track with ample room for the mixer to make the adjustment.

In actuality, the reels of the "Dempsey" TV show, which had major fights, took approximately 18 tracks to cover the action. They were built with punches in tracks 1 through 3, then crowds in tracks 4 through 9, and the balance of the tracks for miscellaneous effects. The purpose of this type of plan is that one dubbing mixer cannot handle 18 tracks at a time. The normal method of dubbing a reel such as this is called predubbing. First, we dub tracks 1 through 3 and get all the punches recorded on one track at the proper level. Then we dub tracks 4 through 9 and get all the crowds properly recorded on one track. Then again we predub the balance of the tracks and get all the miscellaneous effects prerecorded on one track. It is then an easy process to rerecord each of three prerecorded tracks onto a single effects track.

The eventual product of the dubbing stage is a roll of 35 mm magnetic track with three separate channels. One channel takes the dialog, a second the music, and the last is used for sound effects. We can then make changes in effects without disturbing dialog, music, or any combination thereof.

Also, by eliminating the dialog track, leaving only music and effects we create a foreign track (see Chapter 15). This combination of music and effects without dialog is alternately referred to as a musifex track, minus dialog, or any other of several names meaning the same thing. Each foreign language country to which the film is sold then need only to add its voice track to have a complete film.

The building process is one that each editor creates to best fit his method of cutting and other personal idiosyncrasies. You can build one track at a time, two as we have demonstrated, or in my case, I have a six-gang synchronizer and I would build a seven track reel by building tracks 1 through 4 at one time and then tracks 5 through 7 on a second pass through the synchronizer.

SUMMARY

The purpose of building is to place those effects that you have carefully selected and edited into sync with the action into tracks that the dubbing mixer can best deal with, such as similar effects on the same tracks, sufficient room between different effects for the dubbing mixer to change levels and even equalization, no more tracks than necessary (the more tracks to a reel, the more time it takes to get set up for dubbing), and never fewer than required for proper manipulation by the mixer.

Take the time to be sure the effects are cut in sync. Make clean, neat splices that will not rip going through the recording equipment (dummies).

· 11 ·

The Foley Stage

In the early days of sound most action was left silent. Doors opened and closed with no sound. Footsteps were ignored. Over the years and for many reasons the demands on the sound editor have increased. "If you see it, you hear it" is the modern credo. This brings us to the problem of how to cut the sound of a man taking a wad of money from his pocket and counting five bills out and returning the wad to his pocket. This is not done. Instead, sound is made on a Foley stage.

A Foley stage is a room with a screen, a floor with as many as 20 different surfaces, and a microphone. It is a recording studio with a screen and a room full of assorted props. In the back room a roll of mag stock is loaded on a recording machine and put into interlock with the projector. As the film is projected on the screen, the required sounds are created on the stage by the Foley artist (sound editor) and recorded on the mag stock. The same stage is often used for ADR recording (see Chapter 6).

Foley is the process of recording directly to the picture action that is too minute to cut from a library. For example, a man drives up to a house, parks his car, walks up to the door, and rings the bell. A women answers the door as the man reaches into his pocket, counts out five $1 bills, and leaves.

The car drive in, car door open and close, bird background, doorbell, and house door open would all come from a library. The man's footsteps and his counting the money would be done on a Foley stage.

When spotting the show in his moviola the supervising editor will program Foley in much the same way he would program ADR. He notes where footsteps and other minor effects are needed and writes the footage and frame on sheets, called Foley cue sheets (Figure 11.1) along with a description of the action to be Foleyed. Picture is placed on a projector and run to the footage noted on the Foley cue sheets. The Foley artist follows the instructions of the supervising editor and makes the required effects.

The most common effect made on a Foley stage is footsteps. Footsteps can be cut from a library, but it is a long, tedious process. On the Foley stage the walkers watch the action and create the footsteps to match. They have an assortment of floor surfaces available including cement, dirt, sand, gravel, hard wood, plank wood, tile, carpet, asphalt, and grass. If the actor on screen exits a car, walks in the street, then onto grass, and finally onto a cement pathway, the walker duplicates on stage and in sync the action on screen.

Some Foley artists have acquired an amazing capacity for sync. However, a frame of film is 1/24th of a second, and it is impossible to be that precise. It is always necessary to run the recorded Foley in a moviola and cut in into dead sync.

On the stage your imagination can pay dividends. A leather jacket or woman's purse becomes a saddle for leather squeak. A wet towel becomes mud. A roll of magnetic tape serves well for walking in brush. Most Foley stages will have a prop room full of assorted junk. It all comes in handy. Part of an old doorknob and a fork will do for picking a lock or working a safe combination. A kitchen pot sounds just like an army helmet hitting the ground.

The sound effects budget on a show determines the amount of time you can spend on the Foley stage. It likewise determines how exact and precise you can be. For example, the hero is walking down a street. A pedestrian passes going the other way. On a big-budget show you would record the hero's footsteps on one track and the pedestrian on a second track. On a low-budget film you record the two on a single track. The advantage of the big-budget method is in the cutting. Separately, you can cut every footstep putting them into exact sync. Together you are limited. In cutting the hero into sync you may be pulling the pedestrian's step out of sync.

On many big-budget productions every movement is Foleyed. That also comes in handy when dubbing a foreign version. For example, in a recent film it took five Foley tracks to cover the hero: one for his steps, one for the squeak of his leather boots, one for the key chain hanging on his belt, one for his army dog tags, and a fifth for some other part of his costume. That is great when you have the time and budget. On a low-budget show you might have two or at most three tracks for the same

Date: 2-17-86

COMPANY: ___ABC Films_____

MICROPHONES: _____

EQUALIZATION: _____

PRODUCTION: __ABC Production_____

START TIME: _____ FINISH TIME: _____

REEL NO.: ___1_____ SHEET NO.: ___1 of 1_____

MIXER: _____

START STOP FOOTAGE	DESCRIPTION	TRACKS 1	2	3	PROPS
1 START 12 STOP 86.9	Woman walks from cement to dirt.				
2 START 60.8 STOP 86.9	Man walks from grass to dirt to meet woman				
3 START 148.8 STOP 219.5	Scuffle for two men fighting on dirt.				
4 START 171.10 STOP 219.5	Assorted crashes for fight				cardboard box metal tray
5 START 417.11 STOP 602.2	Man and woman walk slowly along dirt path.				
6 START 677.4 STOP 691.5	Man rise from sofa...footsteps carpet and sit on sofa				
7 START 703.12 STOP 818.0	Man walk tennis shoes on cement				
8 START 764.0 STOP	Boy run dirt to cement....stop....exit cement to dirt.				

Figure 11.1 Typical Foley cue sheet.

action. It is your problem to decide how to combine the five elements or which elements to eliminate.

On a big western at MGM we had a scene with two cowboys walking their horses along a trail. We had three tracks of Foley: one for each man's footsteps and one for the rattle of the horse's reins. On a low-budget show we would have put all the footsteps on one track and probably ignored the reins.

Budget determines even the method of recording. On big-budget shows the editors usually record onto a roll of full coat mag using all four channels. Later each channel is transferred to single stripe for editing. For low-budget shows put two rolls of single stripe on two separate machines. Record onto one roll at a time. If there are overlapping Foley needs, simply do one thing, then back up and record the second cue on the second roll of stripe. Although limited to two rather than four channels, no transfer costs are present, which can be high.

A complete, modern Foley stage will also have a sink or water pit. You can fake rowing a boat in a water pit. Try to cut from a library the sound of a man washing his hands. It is easy to duplicate on a Foley stage.

The editors on a big western convinced their producer they had to go to Nova Scotia to make good snow footsteps. After a lot of time and money they came back to Hollywood with nothing. It then occurred to the producer to call MGM and ask how they made the great snow footsteps for *Dr. Zhivago*. The answer was baking soda, big 100-pound bags of baking soda placed on the floor of the MGM Foley stage. When you walked on the bags they crunched just like show. (A lot cheaper than a trip to Nova Scotia.)

For many years MGM had a three-man crew who worked the Foley stage. They were so good many producers brought shows to MGM especially for this crew. One of the men, Kurt Herrenfeld, was great at horses' hooves. He had two small sticks with half a coconut nailed to each end. He would kneel in a dirt pit (the term used to describe the section of the floor with dirt) and recreate the sound, the rhythm, and sync of a horse. A group of horses is easy to cut because there is no real sync but rather a lot of noise. Creating the sound for one horse can be difficult. When watching one horse, you will notice if the hooves are not right. Kurt could match any horse doing anything so well you did not even have to cut them in the moviola.

Almost anyone can learn to become a reasonably good Foley walker. It takes time and experience.

Figure 11.1 shows a typical Foley cue sheet. The first column is for the start and stop footage of each cue. The center is for a description of the action to be covered. There is also a column for listing any props which will be needed.

When programming a reel for the Foley stage the editor must keep in mind the limitations of the stage and the walkers. If a man runs from cement to dirt he should program two separate cues, one for each surface, because it is difficult (if not impossible) for the walker to run from the cement pit to the dirt pit and maintain sync.

SEPARATION

A man walking along a dirt path in the park passes a second man running in place. Properly programmed you will have one cue for the walker and one for the jogger. What we are talking about here is separation. A rule of thumb is the more separation of Foley, the better it can be dealt with in dubbing and also, the more precise you can be in dead sync editing. You have two men jogging along a dirt path. One stops and the other continues running off into the distance. On two tracks the mixers can keep the close-up jogger at a high level while fading the second man off into the distance. On one track you are married to what you record on the Foley stage. On one track you cannot control each man individually.

Unquestionably, the more separation of Foley, the better, but separation takes time and time is money.

We have had some fun on Foley stages. Once my assistant and I collaborated to make a great monster. I worked on one mike making a most distasteful nasal snort. On the other side of the room, with a different mike so the recording mixer could properly balance the two, my assistant breathed a heavy raspy sound. The two blended into one scary monster.

Knowing what to cut from the library and what to Foley comes with experience. Knowing how to use a Foley stage, to do your own walking takes time.

FOLEY STAGE VERSUS LIBRARY

I had an occasion recently to Foley a show for another young editor. We were Foleying telephone set downs, but that is something that should come from a library. The Foley stage should be used to create those effects too specific to be cut from a library. Examples include footsteps, light movement, handling small items such as money, and unwrapping a gift. In other words, sound effects too specific or exact to come from the library should be created on the Foley stage. More common or general effects are easily and more cheaply taken from the library.

· 12 ·

Dubbing

Very few professions provide such immediate and visible tests of the quality of your work as sound effects editing. Whether on your own or with notes from a supervisor, you sit down at the moviola with a reel of film, and have to hear in your mind what it is going to sound like when you have done your job. You cut a car chase effect by effect: car by, car steady, car by, tire skid, etc. How will it sound when all put together? Will the producer like the cars you have used? Does the producer want more skids or fewer? Is there a gunshot on this cut or not? There are an infinite number of ways to cut any given scene. Usually, you will cut it the way you see it, but the producer may not see it the same way. There are no measures by which to judge a reel of sound effects. It is a subjective perception.

EDITOR–MIXER RELATIONSHIP

One of the first feature films I worked on at MGM was a Jimmy Caan vehicle called *Slither*. Prior to dubbing, the sound editors and the dubbing mixer had a check run. The purpose of a check run is to verify all the units and selection of materials before going into a final mix with the producer/director. At the end of my reel Jimmy Caan and his friend were in a do-it-yourself car wash cleaning their motor home. In a fit of happiness Caan turned the spray wand onto his friend, drenching his friend. I had cut a sound for the water hitting the man. The head mixer felt that the effect did not work and told me to change it. Upon my refusal, hard feelings arose. I learned, through this experience, that the dubbing process was not—at that time—considered a partnership between the editors and mixers.

The advent of the independent film and the independent sound company has led to a reassessment of the relationship between the dubbing mixers and the sound editors. The sound editor is now a client. He often has input with the producer as to where the film is to be dubbed. If the sound editor does not like a particular mixer, he will not bring his shows to that company. The relationship, however, is not one-sided.

It is in the power of the dubbing mixer to make the sound editor look good or bad. It therefore behooves the editor to have a good relationship with the mixer. The more properly prepared a show, the easier it is for the mixer to do a good job and the better the sound editor looks.

It is much easier and less traumatic to go into a dubbing session knowing the show is properly prepared, in sync, and laid out so that the mixer can

Figure 12.1 A machine room. Here each individual track of sound the editor has built is loaded onto a reproducing machine. Each machine is wired to controls on the dubbing panel. A projector carrying the picture is put into interlock with each of these machines, commonly called dummies, and the entire system then rolls at exactly 90 feet per minute. The dubbing mixer has exact control of every track for mixing purposes. The system can stop, reverse, and roll forward at normal speed or at speeds six times normal.

Figure 12.2 A dubbing stage in operation. With the film (action) being projected on the screen, the dubbing mixers sit at this panel maintaining precise control of every sound coming to them from the dummy room (machine room). Executives sit behind the panel telling the mixers how they want the various sound elements to be blended.

handle all the elements. Confidence and professionalism are quickly recognized and respected on a dubbing stage.

Understanding the dubbing process is necessary to being a good sound editor. Understanding the mixer's problems will enable you to lay out the reel properly. It is to your benefit to grab every opportunity to sit in during a dubbing session as an apprentice or assistant. If you are not welcome or out of place in the dubbing room itself, stand in the projection room or the dummy room (Figures 12.1 and 12.2).

THE PROCESS

For those totally uninitiated, dubbing is the process whereby all the individual elements of sound are brought together and blended into one smooth track, which the audience hears in the theater. The elements are dialog, music, and sound effects. In turn the dialog consists of the original dialog shot with the picture and any looping done. The music consists of what the composer has written, the orchestra recorded, and the music editor built into reels so that it begins and ends where the composer visualized.

Lastly, the sound effects consist of what the effects editor has built from the library and the results of the Foley stage.

The dubbing crew generally consists of three people: one working with the dialog, one the music, and one the sound effects. As a rule, the head mixer works with the dialog; his is the most important element of the three. Music and effects will always have to give way to make room for the audience to hear the dialog. The dialog mixer's responsibility is to make the dialog smooth, to make any looping sound like the original dialog by equalization, and by reverbing a line of dialog or using any other of the many available tools on a dubbing panel.

The music mixer's job is to balance the music against dialog when both occur simultaneously and to have judgment as to the level a piece of music should be played. The music mixer can make a particular piece of music fade in very quietly and discretely or allow it to bang in like a train. It is the music mixer's job to blend the music into the overall sound track as the composer designed it to be heard.

The sound effects mixer has perhaps the hardest job. On most reels there are many more sound effects tracks to deal with than music and dialog combined. A busy reel of effects may have 40 or 50 tracks. The mixer cannot deal with that many at one time. Generally, as a reel of film goes by on the screen in the dubbing room, the effects mixer's hands will be the busiest.

In a gun battle there are gunshots close-up and at various distances. Riccos, screams, and glass crashes also happen at various distances from the camera. It is the editor's job to have close-up gunshots on one track and distant shots on a different track so that the mixer can deal with them. Still the effects mixer must find the right balance (level) between them. Here is where the partnership between mixer and editor looms important. It is their combined job to satisfy the producer. The editor can only cut the needed material, he cannot create one overall sound track with all the elements required in proper balance to each other. In contrast, the mixer can only deal with the material given him by the editor. If the material is not there or if it is wrong, he is at a loss.

The Dubbing Panel

Physically the dubbing operation consists of three rooms: the projection booth from which the picture emanates; the machine room, a room full of tape recorders (dummies) wherein each track that will go into creating the overall sound for a given reel is mounted onto one recorder, which is in turn wired into a given spot on the dubbing panel; and the dubbing room, composed of the screen on which the film is viewed.

The dubbing panel looks like the control room of a spacecraft. It is approximately 15 feet long with hundreds of buttons, knobs, switches, and slide pots. Basically, it has controls for each tape recorder in the machine room so that the individual mixer can control the sound coming from the recorder. He can control the level, the equalization, he can make it reverberate, or he can make it sound like it is coming over a telephone.

New and Old Dubbing Methods

In the past, dubbing was done on optical track with no forward or reverse. If while dubbing your way down to 900 feet in a reel and something was missed, you had to go back to the head and start all over again. Today with magnetic film, rock and roll dubbing rooms, and the capacity to punch in and out of record it is much easier. It also gives the producer leeway.

Before the advent of rock and roll dubbing rooms, dubbing changes were more costly and time-consuming. If we dubbed our way down to 900 feet in a reel and a mistake was made, *e.g.*, a church bell was played too low, the producer had to make a decision. Was it worth the time to go back to the head of the reel and start all over again just to raise the level of that church bell? Often it was not and the bell was left lower than would have been preferred. Today you just back up the entire dubbing system, punch into record just before the bell, and raise the level. In the old system the change could have taken 1 or 2 hours. Today it can be made in 2 minutes. The advantage is tremendous, and it allows the producer to be more exacting.

Pre-dubbing

A method of dubbing that bears some discussion here is pre-dubbing. Many shows are dubbed as previously discussed, *i.e.*, all dialog, music, and sound effects are played at one time and the overall track is created. Pre-dubbing is also used. It is gaining in popularity. It involves taking one portion of the sound track, usually dialog, and dubbing those tracks. What you then have is a complete, smooth dialog track to which you later start to add sound effects and music. Pre-dubbing has its drawbacks. For instance, many small imperfections in the dialog track which will disappear under music and/or sound effects stand out when heard alone. The dubbing mixer then spends much time overcoming those imperfections that might have been alleviated.

Another argument against pre-dubbing is that after the time has been spent to create a dialog track, it still must be changed to accommodate the music and sound effects. For instance, two people talking at the airport. You dub the scene so that the dialog is clear and audible. Then you add the effects of an airport background including an off scene jet takeoff, and you still have to raise the level of the dialog to get above all the noise. Thus, the pre-dub has to be re-dubbed.

Effects Editor: M. KERNER Producer: ____ Production: DEMPSEY

Track No. 1 Reel No.	Track No. 2 Reel No.	Track No. 3 Reel No.	Track No. 4 Reel No.	Track No. 5 Reel No.	Track No. 7 Reel No.
9 POP					
12 - BIRDS	12 BARNYARD	14 PUNCHES	14 PUNCHES	15 PUNCHES	
				18 BODY FALL	
				73 P/S CHIMNEY	
	187	187	168	169 166	
				172 GORE FALL	
				197 P/S CHICKENS	
			198 DISH BOWN		
			219 DISH BOWN		
				273	
73 MAN YELL - 273 283	PUNCHES - 278 273	PUNCHES - 278	279 BODY FALL	279 CHEER - 280	
88 MAN YELL - 291	292 PUNCHES - 299	292 PUNCHES - 299			
05 P/S DIGGING			305 HORSE BY 320	399 CHEER	
	328 CM BY		336 HORSE BY		
	340			399 BODY FALL	305 MULE BKG
354					
355 PUNCHES	355 PUNCHES		351		357

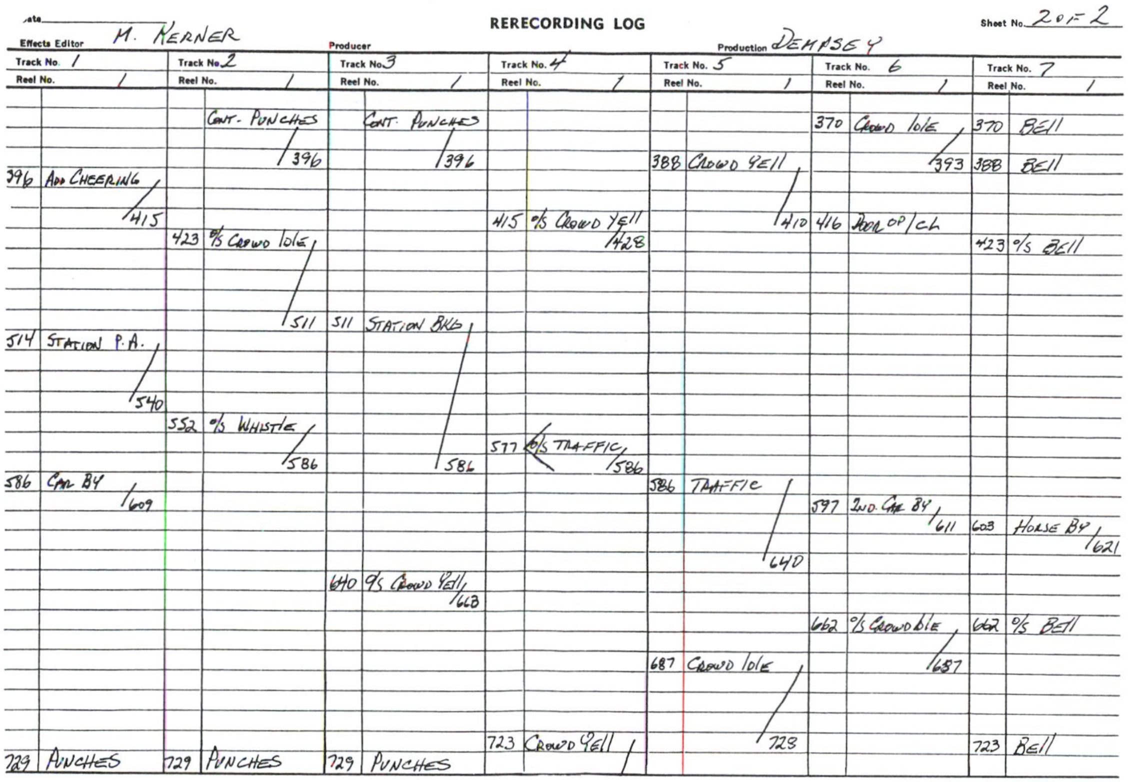

Figure 12.3 Example of a neat, easily read cue sheet, allowing the dubbing mixer to follow easily.

There are, however, advantages to pre-dubbing such as time limitations; for instance, one mixer can be pre-dubbing dialog while the sound effects are still being edited.

PREPARATION

No matter how concerned, thorough, and professional the editor, it is always possible that the producer will want something quickly added or changed during the dubbing session. It is good policy for the editor to bring any extra material he thinks may possibly be usable to the dubbing. In our case we have about 40 film boxes filled with sound effect loops. They run the gamut of possible effects and have often proven useful. For instance, we cut a particular crowd in a bar, but the producer wants it busier or quieter or something different from what we have cut. Among our loops are five or ten bar crowds. We can listen to each, hopefully select one, and continue dubbing within minutes.

Often you may steal an effect from one reel in the film and use it in another. For example, in reel #2 you have a particular door close. You find you need a door close for reel #8. Pull the first door close out of the track in reel #2 and cut it where needed. If the reverse is the case (you need an effect in reel #2 that has been cut into reel #8), take it from the track in reel #8, but as soon as reel #2 has been completed put it back into #8.

CUE SHEETS

When making out dubbing cue sheets there are several things to remember. The dubbing mixer sits in a semi-dark room, with the film running at 90 feet per minute. He must be able to follow your cue sheets while moving his hands about from pot to pot adjusting levels and equalizations. It is the primary purpose of your cue sheet to allow that mixer to read the footages and effects quickly and easily. The notes should be kept simple. Do not go into lengthy descriptions of each effect (Figure 12.3).

It also helps the mixer if you keep an even flow to the footages. If you examine the sample dubbing cue sheets on the following pages, you will note the footages work evenly as you go down the page.

It is helpful to consider each line as 10 feet. Therefore, an effect that runs 50 feet will take five lines before you write the ending footage.

Note the < at 577 feet in track #4 in Figure 12.1. That indication informs the mixer to fade in the effect. The same mark reversed at the out footage of an effect would require fading out the effect.

The purpose of indicating o/s at the beginning of an effect is to inform the mixer that the effect is to be played lower than normal to create a distant quality to the sound.

All similar effects should be kept in the same track. For instance, all bells are in track #7, punches are in tracks 3, 4, and 5.

Once the mixer has found the proper level and equalization for an effect, it is easier to leave those controls set for the next time the same effects come up.

You must find a reasonable compromise when making cue sheets. You do not want effects too close together, thus hindering reading. You also do not want more pages than necessary. It is time-consuming and distracting for the otherwise busy mixer to reach up and move a sheet aside.

Like every other set up in sound editing the best way to learn how to make out a proper dubbing cue sheet is to watch others and to ask the dubbing mixers how they prefer to have their cue sheets made. Simplicity, neatness, and accuracy are the goals to effective cue sheet preparation.

THE CHANGE ROOM

Adjacent to every dubbing stage is a change room, a fully equipped editing room made available to the visiting sound editor to make additions, corrections, and repairs (Figure 12.4). The better prepared the show, the less

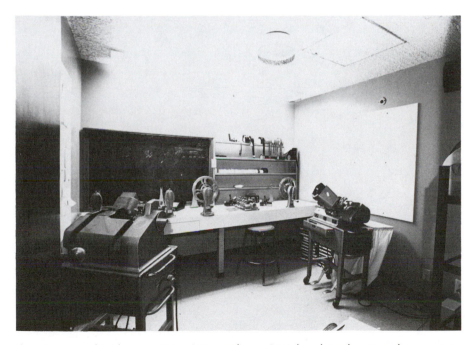

Figure 12.4 The change room. A properly equipped and ready-to-go change room attached to a dubbing stage. This is where you try to make the requested changes or required repairs to tracks during a dubbing session.

time will be spent in the change room. When working in the change room you generally have a stage full of people waiting for you; the pressure is on. Although speed in making the change or repair is needed, accuracy is also a prime consideration. Making the change correctly alleviates possible complications.

MIXING THE GOOD WITH THE BAD

A dubbing session offers a chance for creative input. The better editing job you do in the cutting room, the less likely these problems are to arise. As a result, the change room may not even be needed.

A good mix is exciting. You see all your work come together and contribute to making a film come alive. The dubbing room is where you and your work are judged. The quality of your work reflects on you personally, and it can garner (or conversely lose) respect. A properly prepared reel, in sync, with a reasonable amount of added atmosphere effects, and carefully chosen effects that work indicate your professionalism. That is how a sound editor becomes part of the creative process of filmmaking.

Finally, it is important to learn your trade well and to take a sincere interest in the film and in the producer's welfare.

· 13 ·

The Library

The sound effects library is a catalogued collection of sound effects. It consists of almost every imaginable sound from autos to wind. These sounds can live in the form of quarter-inch tape or 35 mm magnetic tape. Each has its advantages and disadvantages.

ORIGINS OF SOUND EFFECTS LIBRARIES

The origins of sound effects libraries go back to the major studios at the beginning of sound. In the late 1920s and early 1930s the studio sound departments of the major studios had tremendous power. When the sound effects editor needed sound effects the editor simply went out and recorded them. Even in later years this often held true. I once worked on a TV series called "Then Came Bronson." Before we started the first show the studio hired a motorcycle stunt driver, bought three bikes, and sent us all out with a production recorder to create our own library of motorcycle effects. We spent an entire day at a local race track getting all the sounds we would need for the series. Sound effects libraries can be continually added to and improved.

AN IMPORTANT TOOL

The library is the most important tool any sound editor has. Without the right material he cannot do his job well. A poorly organized library will cost time searching for material. A library with worthless material is even worse because the editor will spend time pulling and running unusable material. Thus, an efficient library should have good, usable, and well-catalogued effects.

Our library is relatively small; it is composed of 500 film boxes with an average of ten effects per box. The material is well organized and all usable.

Suppose you need the effect of unscrewing a light bulb from a socket. A poorly organized library might have two effects with two separate numbers: one for screwing the light bulb in and the other unscrewing. It would be more efficient to have both on one master with one number. It would reduce storage space, catalogue space, and time.

LIBRARY ORGANIZATION

There are an infinite number of ways to organize a library. Figure 13.1 provides excerpts from a sound effects library. Should you ever have the occasion to either start or reorganize a library, it is suggested that you do so in a manner that makes sense to you. Following is a breakdown of how our library is organized:

Autos: categorized by car, then by what each car is doing.

Auto misc.: such items as doors, trunks, skids, handbreaks, windows, windshield wipers, *etc.* Any sound having to do with a car other than the engine.

Airplanes: subcategorized by jet, prop, and helicopter.

Animals

Boats

Break: anything being broken such as a tree branch, an egg, eyeglasses, a door, a film can, *etc.*

Backgrounds: subcategories are interior and exterior, such things as harbor, park, barnyard, race track, airport, rail station, desert, construction site, *etc.*

Bells

Birds: subcategorized as individual such as thrush, seagull, owl, or crow and bird backgrounds such as jungle, woods, pigeons, and tropical.

Clicks

Crowds: subcategories include applause, babble, boo, cheer, courtroom, children, excited, foreign, laughter, men, party, murmur, reactions, yelling, women, and places.

Crash

Clock

Cameras

Coins

Doors: subcategories include wood, metal, glass, sliding, elevator, and then miscellaneous to include door knocks, bells, buzzers, and key sounds.

Elevators: (It is often worth the effort to cross-file some effects. We did here with elevator doors filed under both doors and elevator.)

Electric

Fight: including socks, body punches, kicks, body falls, slaps, *etc.*

Fire

Falling: Anything that falls to the ground, *e.g.*, tin cans, suitcases, a bundle of newspapers, a ladder, a briefcase, *etc.*

Footsteps: subcategories include individual and group.

Glass: breaking windows, plates, bottles, *etc.*

Guns: subcategories include pistol, rifle, machine gun, artillery, riccos, clicks, and miscellaneous.

Gambling: any sound having to do with gambling, such as roulette wheels, dice, cards, and slot machines. Here we also cross-file gambling crowds.

Hits: subcategories include metal, glass, wood, and miscellaneous.

Horses' hooves: subcategorized by single, 2, 3, and posse, then by pace.

Horns: includes boat, train, and miscellaneous. (Automobile horns are categorized under auto misc.)

Kitchen: anything found in a kitchen, such as an electric knife, can opener, garbage disposal, steam kettle, toaster, *etc.*

Machinery: a large category broken down by home, business, construction, hospital, industry, and miscellaneous.

Motorcycles: arranged in much the same way as automobiles.

Miscellaneous: some items defy categorizing and are just thrown into this category.

Paper

Rolling

Sirens

Scratch-scrape-squeak: includes many hard to define sounds such as a drawbridge squeak, sharpening a knife on a whetstone, erasing a blackboard, metal stress, lighting a match, *etc.*

1979 OLDSMOBILE

```
1- Start----idle---Revv----idle stdy @ 10, 20, 30, 40, 50--------G24
2- Start--idle--Revv--idle--out--in--off @10 mph --2X------------G24
3- same at 20MPH---------------------------------------------------G25
4- same at 30MPH---------------------------------------------------G25
5- same Fast-----2 outs --1 in------------------------------------G25
6- Start---idle---Revv---out fast--slight skid--------------------G25
7- BY-----@10-----2X----------------------------------------------G25
8- BY-----@20-----2X----------------------------------------------G25
9- BY-----@30-----2X----------------------------------------------G24
10- BY----@50-----2X-----VG---------------------------------------G26

11- By-----Fast-----VG--------------------------------------------G26
12- Start--rverse from mike---out by slow-----------------------G26
13- same----normal speed------------------------------------------G26
14- same----fast speed----with skid------------------------------G26
15- Start---reverse to mike---out---slow-------------------------G26
16- same-----medium speed-----------------------------------------G26
17- same-----fast speed-------------------------------------------G26
18- in----idle at signal--out @20--@30--@40----------------------G26
19- Maneuver-----mostly dirt---to and fro---do all--------------G27
20- Maneuver-----dirt---good cement on end---VG------------------G27

21- Passby------20 with horn---30 with horn---------------------G27
22- Passby------fast-----with horn------------------------------G27
```

Accessories

```
1334- Horn-------do-all-----VG-----------------------------------G27
1335- Doors-----------------------------------------------------CD1
1336- Hood---int. release---ext. close--------------------------Z80
1337- Trunk----open with key---close----------------------------Z80
1338- Gas Cap-------do-all--------------------------------------Z80
1339- Electric windows-----------------------------------------Z83
1340- Emergency brake---foot set---hand release-----------------Z83
1341- Gear Shift----column mount---auto trans.------------------Z85
1342- Heater-----on----stdy--------off-------------------------Z85
1343- Air Conditioning----on--stdy---off-----------------------Z85
1344- Keys----turn of ignition---or other????------------------K71
```

(A)

Figure 13.1 Excerpts from a sound effects library. (A) The first example shows complete coverage of a 1979 Oldsmobile. Items #1 through #22 include all engine sounds, whereas #1334 through #1344 include assorted sounds related to an automobile. The digits to the right indicate where the master effect is located. (B) The second example is a 175 horsepower outboard motorboat. We attempted to record all the possible sounds of the boat including starting, leaving, passing, constant speed, coming in from a distance, and turning off. For proper coverage we need all these actions recorded at

175 H.P. MERCURY OUTBOARD

```
1 - INT. MIKE... idle-out to medium speed, steady............... Z57
2 - SAME - slightly faster....................................... Z57
3 - SAME - fastest (fludder-accelerate)......................... Z57
4 - INT. MIKE... start, rev, off................................ Z57
5 - INT. MIKE... starting out hi speed, maneuvering variable
                speed, in, off.................................. Z58
6 - INT. MIKE... false start, start, rev, idle, off (closer mike)..Z58
7 - INT. MIKE... starting out to cruising speed (20-25m.p.h.)
                variable speed, long slow in, long idle, off... Z58
8 - START, away, slow, short.................................... Z58
9 - START, away, med. speed, long.............................. Z59
10- START, away, med. speed, (BEST)............................ Z59
11- START, away, fast, short................................... Z59
12- PASS-BY, med. speed........................................ Z59
14- CIRCLE MIKE.. var. speed, heavy load, water bg............. Z59
15- PASS-BY, med. speed, v.g. approach, short away, pan mike.... Z60
16- INT. MIKE.. false start, starting out, med.-hi-speed maneuver,
   (300 ft.) slow down, speed up, slow speed rev, idle, off (DO ALL). Z61
17- med. speed, mike over side, variable speed.................. Z60
18- PASS-BY, med. speed........................................ Z60
19- PASS-BY, hi-speed.......................................... Z61
20- PASS-BY, var. hi-speed..................................... Z61
21- PASS-BY, hi-speed, pan mike V.G............................ Z60
22- PASS-BY, hi-speed, pan mike V.V.G.......................... Z60
23- PASS-BY, med. speed, stat. mike............................ Z60
24- PASS-BY, long, stat. mike.................................. Z60
25- PASS-BY, med. speed (good) short away...................... Z62
26- PASS-BY, hi-speed, longest, V.G............................ Z62
27- PASS-BY, hi-speed, best approach, pan mike, V.G............ Z62
```

(B)

various speeds. (C) The third example shows but a few of the different types of traffic one would have in a complete sound effects library. There are foreign traffics with a definite flavor such as Rome, which has many scooters, and Amsterdam, which still has trams. Small town traffic is different from New York City traffic. Older traffic might have the sound of Model A, whereas modern traffic is different. Traffic to be used at a major intersection is different from traffic to be cut for a small side street. Traffic on Sunset Boulevard in Hollywood travels much slower than traffic on an interstate highway.

TRAFFIC--------Pg#2

```
 158- London-----outside Hilton Hotel-----------------V2
 315- Amsterdam----birds and trams--------------------V3
 316- Amsterdam----outside Krasnapolsky Hotel----------V2
 442- Traffic-----------------------------------------V2
 722- New York City------big--------------------------Z74

1123- Passbys------wet-------------------------------Z3
1168- Traffic Jam horns---(Auto#)----big-------------X41
1431- Traffic Jam Horns---(auto#)--------------------Z18
1757- Distant-----indescript for fill only-----------L1
1801- Freeway-----from one block away----------------T10

1825- Lot #3---MGM----fill use only------------------Z46
1802- Older----streetcars and bells------------------Z95
1872- Sunset Blvd.-------VG--------------------------L1
1981- New York City-----VG----horns-----------------K56
1985- similar to #17---horns--VG---1000'-------------Z54

1986- New york City-----VG----big----horns-----------X12
4301- light----from interior at night----------------G23
4302- WET-----big city-------------------------------T12
5183- Small Town-----distant activity----------------G9
5329- Highway--------VG----cars by fast--------------X13

4315- New York City----o/s Dock area-----------------G23
6626- Big City----distant---birds---jet by-----------G19
HK23- Hong Kong----big----some horns-----------------H16
HK24- Hong Kong----heavy--no voices------------------H16

   34- Rome----with voices---from street cafe---------Z98
   35- similar to above--less voice--CU Bys-----------Z98
```

(C)

Figure 13.1 (continued)

Sport

Space

Traffic

Telephones

Trains

Thunder

Voice: subcategories are men, women, and children.

Water: subcategories include fountain, falls, splash, river, running, rain, surf, and miscellaneous.

Wind

Windows

Wagons

Whistle

Working

CARE AND MAINTENANCE

Of vital importance is the care and nurturing of the library. In the past the major studios had librarians whose full-time job was maintaining the library. Today few studios, if any, spend the money necessary to properly maintain the library. As a result, many are in disarray. When your editor needs an effect, pull the master, make a copy, and then put the master back where it came from. When refiling a master it is vital it be filed correctly. If a master belongs in and is catalogued in box #200 and it is misfiled into #346, it is probably lost. You might as well have thrown it away.

Every time a master is run over a sound head there is some loss of quality. Eventually, a popular master will wear out. It is imperative someone watch for this and make a new master before the original is too far gone. There are many ways a library can grow such as by recording effects yourself, trading with other companies, and producing effects recorded with the films you edit. Sound effect records on the market have some excellent effects that would enhance any library. As long as each and every effect is usable, *i.e.*, clean of background noise and a quality recording, it is worth having in the library. The library is the tool of the trade and should be treated with the greatest care possible.

LIBRARY FORMATS

Some libraries are kept on quarter-inch tape. It is a good way of storing a large library in a small space. Other libraries are on 35 mm magnetic tape. Each has its own advantages.

One advantage of 35 mm is loops. We have a bird track we use all the time for general all purpose, backyard birds. We use thousands of feet of this particular bird. We keep a 40-foot 35 mm loop master. When we need footage we run 1,000 feet of the loop in 10 minutes. If our library was on quarter-inch we would have to go back and forth and make probably ten prints of 100 feet each and then splice them together to get a 1,000-foot reel of the birds. The overall cost of that 1,000 feet would be five times greater than with 35 mm. Another major advantage of 35 mm masters is the ability to run a master against the picture to ensure the effect will work before transferring, which can save time and transfer costs. However, 35 mm magnetic tape can be eventually worn out, and it takes up much storage space in a library.

· 14 ·

The Assistant

Much of what a sound effects editor does can be done by an assistant at a much lower cost to management. A good assistant can double the output of the editor with whom he works. The main function of an assistant is to take the cutting notes and turn them into kits, thus providing the editor with a box containing all the material needed to complete a reel.

Referring once again to our cutting notes, in Figure 10.1 the assistant must give the editor a print of birds #296, background #1241, and fight #819, #823, and #824 as well as all the other effects listed. How much material is needed will also be considered. For example, birds #296 have to run from 12 to 187 feet. The editor should have at least a 200-foot print of birds #296. Animal #99, a horse whinny, is at 73 feet. Obviously, a whinny will be 3 or 4 feet long; therefore, the editor needs a print of only 20 or 30 feet to be certain of having the needed whinny.

In many cases the assistant will build the reel for the editor. If all the impact effects are properly marked and the roll-in effects sufficiently indicated on the picture, there is no reason the assistant cannot build the reel while the editor goes on to cut the next reel.

The cutting notes form shown in Figure 10.1 was also devised here and has shown itself to be functional. The form allows you to note the starting and ending footage of an effect plus the track number into which it is cut and, when desired, the library number of the effect. It is from these notes that the dubbing cue sheet is made. The form is additionally valuable in the dubbing room because it allows you to follow the action, being certain every effect is played. Many editors go into the dubbing room without notes and, when asked a question about the reel, must refer to the cue sheet. With the cutting notes he has all the information about his reel at his fingertips. The assistant can make out the dubbing cue sheet from these cutting notes.

Many studios and independents have laid the responsibility for the library on the shoulders of the assistant. Once the province of the librarian, maintaining the library can take much of the assistant's time.

While keeping up with all these responsibilities, the assistant is supposed to learn to become an editor. The ambitious assistant will take every opportunity to learn from the editor. The more knowledgeable the assistant, the more help he can be. Helping the editor spot a reel, then pulling the effects, building and watching the reel dub gives the assistant a wonderful overall view of sound editing.

Before graduating to assistant, many people get their start as apprentices, either the editor's guild, if it is a union job, or in a nonunion editing room.

Terminology, lab work, relationships between various members of the post-production team including the film editor, assistant editor, sounds effects editor, music editor, negative cutters, lab men, *etc.*, all are part of the system through which a film must pass. The more you know about that system, the better your chance of advancing through the ranks. In film as in any other industry, your superior will react positively to your desire to learn and willingness to do whatever is needed.

Often, cutting Foley is a good opportunity to get used to working with a moviola. Again, a willingness and enthusiasm to do anything and everything asked or offered is your best approach to learning sound effects editing.

· 15 ·

Foreign Language Films

Foreign sales of American made, English language films is an important source of income to the producer. For that reason almost every film requires a foreign version to be created. Other terms for the foreign are M&E (music and effects) and minus dialog. By any name a foreign is an American made, English language film with a complete music and sound effect track and no dialog. Each country the film is subsequently sold to then adds its own native language to the dialog track.

There are two ways a foreign can be created. The first is from a normal dub, *i.e.*, one track of 35 mm film with the dialog, music, and sound effects each on its own channel within that 35 mm film. Another way is from composites.

FOREIGN FILMS FROM SEPARATION DUBS

By far foreign films from separation dubs are the easier of the two approaches to foreigns. The editor can make a copy of the sound effect channel of the original mix, run it against the film, and easily and rapidly see what is missing.

In the recording of the original production sound there will be many sounds that do not have to be cut or recreated by the sound editor. For instance, two people walking down the street have their footsteps recorded along with their dialog. The sound editor does not have to Foley their steps. Two people are having lunch; as their dialog is recorded we also get the sound of their cutting a steak or setting down a glass on the table. We do not have to recreate those sounds. However, when we eliminate the dialog track entirely these sounds are lost. They are the sounds that must be created for the foreign, original production sounds that are lost under the English dialog.

A good approach is twofold. (1) Run the sound effect channel of the mix against the picture and make notes on every missing effect. Then run a copy of the dialog channel against the picture to see if any of the previously missing effects are on the dialog channel and clear of actual speaking. They can then be saved. (2) Move any good effects from the dialog track to an effect unit during the original editing process. Those effects will then be contained indefinitely on the effects channel.

Having run both the effects channel to see what is missing and the dialog channel to see what can be saved, make specific notes on what must be done. The required effects are cut, Foley recorder, *etc.*

In the dubbing of the foreign we take all the sound effects from the original mix right off the master of that mix to eliminate losing more generations on the transfer of the sound. The additional sound effects are added as well as Foley and the effects from the saved dialog track. The result is a complete sound track minus dialog.

FOREIGN FILMS FROM COMPOSITES

Occasionally a producer has a film for which there is only a single composite sound track, *i.e.*, one track with sound effects, dialog, and music combined. The producer needs to make a proper foreign. This is more complicated. You must take a transfer of the sound track and first eliminate all dialog. That will leave big holes in music and sound effects. Your job is to fill those holes. Here we deal only with the problems arising from missing or incomplete sound effects.

Using as much of the original material as you are able to rescue, you then refer to a sound effects library to complete other scenes. A barroom fight is going on. Midway through there is some English dialog. By making a copy of the first half of the fight, we can repeat that to cover the second half of the fight that is no good because of the dialog. You may then have to cut a few specific close-up shots or body falls, but the main body of the fight and crowd yelling will be there.

For a film buff or producer this problem is of little interest, but as a working sound editor it must be resolved. Why any film dubbed in the last 15 years or so would have only a composite track seems ridiculous. Perhaps there was a normal three or four channel separation track made and subsequently lost.

· 16 ·

Dirty Harry:
A Critique

An excellent method of teaching sound effects to the novice is to analyze a well-known film. At first I thought of using a classic such as *Casablanca*, but because the art of sound editing has advanced since that movie was made, I opted for a more modern film. In my opinion the best use of sound effects in film today is by Clint Eastwood. To this end I rented a videocassette of *Dirty Harry* and made rather extensive notes on the entire film. (To properly follow my analysis, I suggest the reader do the same.)

For those not familiar with the plot sequence, Dirty Harry Callahan (Clint Eastwood), a San Francisco cop, is called in to save the city from a serial killer, who threatens to continue killing unless paid a large ransom. Scorpio, the killer, says he has buried a teenage girl somewhere in the city and will let her die if his demands are not met. Harry's job is to deliver the ransom money. He is led throughout the city by phone calls from the killer until finally Harry meets up with Scorpio in a park. Once he gets the money, Scorpio knocks Harry down and begins kicking him. Harry's partner runs out of the bushes where he has been hiding and is shot. Harry knifes the killer in the leg causing him to limp painfully away. Harry chases the killer to Kezar Stadium, a football field, catches him, and tortures him until he reveals the location of the buried girl. It is discovered, however, that the girl is already dead.

After the killer is released from custody on a legal technicality, he tries to frame Harry with a police brutality charge by hiring someone to beat him up and then pinning the blame on Harry.

The killer's next target is a school bus full of children, which he hijacks for ransom. Harry has had enough; he goes after the killer on his own and an elaborate scene aboard the bus is followed by an on-foot chase in a factory. The killer manages to slip away to the bay where he grabs a young boy and puts a gun to his head, threatening to shoot unless Harry drops his gun. Harry shoots the killer anyway, injuring him. The famous sequence follows in which the killer eyes his gun and thinks Harry is out of bullets. Harry eggs the killer on with his immortal words (". . . do (you) feel lucky? Well, do ya', punk?"), and when the killer goes for the gun Harry blows him away. The cops arrive as Harry, disgusted, tosses his badge into the bay.

Film by nature and most especially film editing and sound effects editing is a very subjective art. There are no rules, no rights, and no wrongs. It must also be emphasized that what is in the film does not necessarily reflect on the sound editor.

The bold and unique use of sound effects in all of Clint Eastwood's films reflects the personal preference and creativity of the producer/director (in this case Don Siegal). Although the sound editors employed in his films are unquestionably top-notch, the use of effects is most assuredly attributed directly to Clint Eastwood or whoever is the producer/director. For example, during the rooftop scene when we hear a voice over the bullhorn before we hear the sound of the helicopter from which it is coming, I am sure the editor had the sound of the helicopter running well ahead of the voice. It was probably Don Siegal who instructed the dubbing mixer to wait to use the copter sound until after the voice. That is the type of creative control the producer and/or director maintains on the dubbing stage. It is the job of the good sound editor to provide all the material the producer and/or director need to exert that creativity.

The music, incidentally, was by Lalo Schifrin (and as usual was excellent). The better the score, the more likely the producer is to keep unneeded sound effects from getting in the way of the music.

The opening of the film is mysterious. The score carries the action and effects are avoided. The only sounds we hear are the rifle shot and the sound of the girl drowning with a good quality to her voice as she slides underwater.

In the mayor's office we stay pretty clean. A clever exception was bringing in the quality of San Francisco with a trolley bell. The bell helps keep the office from feeling stagy and places us in San Francisco by using a sound easily recognizable.

You will notice backgrounds are generally avoided by Eastwood. Throughout the film the police station, building lobbies, even street scenes and train stations are played very quiet. It is an approach taken for his reasons. There are times I might prefer some feeling of activity and people, which are lacking in this particular film.

The first action, a bank robbery, is played straight, with the usual scream of pedestrians, car crash, fire hydrant, *etc.* The one major difference is Harry's pistol. He establishes immediately a giant sound to the gun. It is exciting and unique. It becomes his signature. It gives him stature and commands fear as well as respect.

The helicopter passby as the police patrol the skies over San Francisco is good, with long approach and good doppler passby quality. The level (volume) of the helicopter when we cut close-up for Harry's exit is too low. It would have added to the excitement somewhat had they played the copter louder on the close-up and also on the following shot where we are flying about inside the copter.

The second time we are on a rooftop with the killer all the ambient sounds are again played (traffic, *etc.*) very low to make room for the music, which carries the scene better than some boring traffic noise would. Again, this is a very creative touch. Both the audience and the killer hear the cop's voice from the helicopter before we hear the sound of the copter itself. The killer might reasonably have reacted to a helicopter approaching, but the use of the voice to break the mood was more shocking and incisive.

In the scene with the potential suicide, I was surprised by the lack of any engine sound for the hoist lifting Harry up to the rooftop. It was a void in the sound of the scene and would have added to its reality. Other than that void we see an excellent example of building a scene with sound. The crowd, sirens, police whistles, and wild lines of specific dialog keep it alive and real.

The sequence with Harry and his partner staked out on a rooftop looking for the killer has a low squeaking of the "Jesus Saves" neon sign. Also, the various electric sounds of the sign being shot out were really first-rate. Here again the scene was kept clean by avoiding much use of traffic or other ambient sounds, allowing the effects they did use to come through cleanly and clearly, thus accomplishing the desired effect.

When the killer calls Harry on the pay phone while Harry's partner monitors the call on the police car radio, there was a well-done split of the killer's voice to create a difference in the quality of the phone voice between Harry's location and the car. The editor properly took the time to have the voice on two different tracks so the dubbing mixer could create one degree of futz for the phone and another for the radio.

The entire scene of Harry being shuttled about San Francisco at the killer's direction was well done. All effects were kept minimal except the important ones. The gun slicks and subway sounds were very good. The lack of traffic, harbor sounds, and subway station backgrounds helped create the suspense. It left Harry on his own, separated from the rest of the city.

The night scene confronting the killer at the cross was so quiet even the crickets were given the night off. Crickets tend to warm a scene, whereas silence is very lonely. The Foley of the killer hopping about after being knifed in the knee was well done. It is a small thing but not easy to do.

In the bus scene I liked the way they trailed off the sounds of the kids singing long after the bus motor was gone. It identified us with the kids and added dramatic value.

All the effects of the bus chase, the car sideswiping, etc., were very good, but I especially appreciated the way they handled the sound of the rock conveyor belt after the bus crash. The plant was first introduced with a long establishing shot, no conveyor. Then we cut to a medium long shot and introduced the conveyor very low. As we cut into the plant itself, the conveyor was raised to a medium level. The effects came to full level as we cut onto the belt, which worked well. As we left the belt, the levels of the conveyor belt again went down in logical steps. A good effect properly cut and well mixed helped build the emotion of the scene. That is what effects are all about.

After Harry shoots the killer I liked the very long shot track of the siren approaching from a great distance. It informed the audience that Harry had done the job and now the conventional police were coming on the scene.

Overall, I would rate *Dirty Harry* as a top flight sound effects editing job and a very good dubbing job. With the exception of a slight disagreement on the lack of a few backgrounds and the low level use of a couple of others, I very much enjoyed the use of sound effects in *Dirty Harry*. I would recommend a study of the film by any reader interested in understanding the creative use of sound effects in motion pictures. However, it is advised to make the effort to remain outside the story and concentrate only on the sound effects.

· 17 ·

Summary

I hope the reader has gained an insight (1) to the importance of sound effects in film and (2) to the resulting importance of the sound editor to the finished film.

If the reader has absorbed some comprehension of the physical act of the actual editing process, all the better.

If I were to include in this work a complete set of notes on a given film the reader would be impressed. To take a show from the time a film editor has completed the editing process and follow it through to the completed dub is a long and laborious process. It requires hundreds of notes and hundreds (if not thousands) of splices. The average ten-reel film might require from 10–40 fixes on each reel of dialog track. Once you have gotten rid of all the director's voices, fixed the clipped words, and smoothed the background ambiences, you must still cut the many sound effects. If your project is a typical TV movie of the week, the required effects will be rather simple. Traffic backgrounds, birds chirping in the yard, a few door closes, and telephone rings will get you through most reels.

If, however, your film is an action opus, you have a distinctly different proposition. I did a TV movie called "Stagecoach," a remake of the old John Ford classic. One reel had over 200 gun and rifle shots, a stage coach at full speed, a tribe of Indians hooting and hollering, arrow zings, bow twangs, riccos, arrows into wood, grunts, horses' hooves galloping, whip cracks, and so on. That one reel had more effects than many entire TV movies.

In the dubbing room there is often a tug-of-war between music and effects. The composer's usual answer is "did you ever see anyone leave a theater humming sound effects?" It has never been my desire to denigrate the importance of music. Equally I resist any effort to do the same to sound effects; the two are compatible. There are times when music only works best for the film and any effects should be left out. Conversely, there are times when the effects sell the action and music just muddles the sound track without adding anything. The sole measure should be what works best for the film, and that subjective opinion ultimately lies with the producer.

As a sound editor, you have an opportunity to add your own touch of artistry, your own creativity. If you set high standards for yourself, they will reflect in your work and eventually in your professional success. A bull painted on a velvet cloth hanging in a Tijuana tourist shop is art. Picasso is art. Sound effects is art. . . . It is up to the editor which kind.

The ultimate desire of a producer is to make the best film possible. That requires not only a good script, actors, and director, but a good sound editor. To ignore the needs of the sound editor in terms of time alotted or budget is to subtract from a film. The original budget should allow for a reasonable sound editing job. The producer should hire the best sound editor available, one who will take a sincere, personal interest in the film. The producer should demand the best from the editor and take an interest in the editor's work. As discussed earlier, film is the only art form requiring the bringing together of many talents. It is the producer's job to find those talents, pay them well, create an atmosphere of creativity, and demand the best.

Glossary

Academy Term referring to start mark appearing 12 feet before first frame of picture.

Automated dialog replacement (ADR) An automated system for replacing original dialog. Rerecording dialog.

Ambience Natural background noise. It can refer to waves at the beach, traffic on a downtown street, or the ambience in a room.

Babble Usually used in reference to a crowd babble. Active voices with no discernible words. More active than murmur or walla.

Background track An all-encompassing effect of ambience for a given location. Samples include a train station, voices and dish noise in a cafe, auto plant, or bowling alley.

Build The act of splicing cut effects into a track so that the sound comes at the desired footage.

Change room Fully equipped editing room immediately adjacent to dubbing room where editor can make last minute changes or additions to tracks during dubbing.

Code A series of letters and numbers printed along the edge of the picture and track used to keep them in sync.

Cue sheet A form listing all the effects, the footage at which they appear, and the track in which they appear; used by dubbing mixer.

Dailies Output of picture and sound from a day of shooting.

Dialog track The sound recorded with picture on the stage or location.

Doppler Term referring to change in the quality of sound approaching and leaving as occurs when an airplane flies overhead.

Dubbing The process of blending all the individual sounds of dialog, music, and effects into one track each in proper relationship to one other.

Dummy A machine that reproduces the sound on one track. The dummy relays that sound through the dubbing panel where it is treated and relayed back onto the magnetic track, which composes the final sound for the film.

Dupe A copy of the color work picture; usually black and white made for the sound editors and music editors.

Fade in Causing a sound to go from zero level to full level at the proper speed.

Fade out Causing a sound to go away naturally.

Foley Recording sound directly to the action of the picture.

Foreign A complete sound track of music and effects with no dialog.

Full coat A piece of 35 mm film coated from edge to edge with emulsion that accepts sound.

Futz The quality added to sound that is coming over a radio or telephone.

Kit A box containing all the material needed to edit one reel.

Level A term referring to volume.

Lavaliere mike A small microphone pinned to the lapel or hanging about the neck of an actor.

Library An organized and catalogued collection of sound effects.

Loop Any piece of film with its two ends spliced together. Rerecorded dialog.

Leader Any type of blank film into which magnetic sound track is spliced.

Mixer A trained specialist who blends all the individual sounds into one harmonious track.

Modulation Sound.

Moviola A machine that runs picture or track or both in sync.

Murmur Low, imperceptible voices; as in a crowd murmur in which no specific words exist.

Negative The film that comes out of the camera from which all copies are made. Any master copy; as in a sound effects library.

One–Ones (1–1s)   An exact copy of the dialog track as composed by the film editor; usually made as protection before sound editors begin their work.

Optical   A narrow band of light and dark striations that pass under a beam of light and reproduce sound.

Production reprint   A copy of the sound recorded for a given scene.

Production track   The sound recorded on the stage that the film editor edits in sync with the film.

Reel   A metal or plastic form onto which film is wound. Common term used referring to a segment of a film, usually 10 minutes.

Reverb   Echo; whether natural as in a recording made in a prison cell or added mechanically on the dubbing panel.

Ricco   Ricochet.

Rock and rolls dubbing   Being able to stop the projector and all inter-locked tracks, back up, and go forward again going into record at will.

Splice   *Hot splice* is to take two edges of a film, scrape the emulsion off, apply glue, and press together, causing them to become one. *Tape splice* is to cut two exact edges of two pieces of film and apply a small piece of splicing tape, causing the same result.

Spot   Running a film a while making notes on every bit of work required.

Stripe   A piece of 35 mm film with only a small portion coated with sound reproducing emulsion.

Synchronizer   A machine that allows two or more reels of film to be run at exactly the same speed; as in keeping picture and sound in sync.

Sync   The matching of sound to action.

Trim bin   A metal bin on rollers with an elevated rod containing hooks onto which an editor may hang film for later use.

Walla   Similar to murmur but more active.

Wild line   A line of dialog with no visable sync.

Wipe   Eliminating the emulsion from a piece of track causing the sound to decrease gradually.

Work picture   The first color print of a film used by the film editor. The negative is cut to match the work picture.

Index

Academy of Motion Picture Arts
    and Sciences, 4
Actors, 39
Astaire, Fred, 39
Automated dialog replacement
    (ADR). *See* Logging

Babble, 42
Backgrounds, 43
Building process, 61–65
Bullitt, 20

Caan, James, 73–74
Change room, 81–82
Coding, 28
Composite sound track, 94
Crowds, 42
Cue sheets, 80–81
Curtiss, Ken, 39
Cutting, 51–59

"Dempsey," 12, 23–26, 65
Dialog track, 27–32
Dirty Harry, 95–98

Dr. Zhivago, 70
Doppler effect, 55
Douglas, Kirk, 8
Dubbing mixer-sound editor
    relationship, 73–75
Dubbing panel, 76–77
Dubbing process, 73–82

Eastwood, Clint, 95. *See also Dirty
    Harry*

Faking, 43–45
Filling, 29
Film apprentice, 2–3
Flatbed editing system, 18
Foley stage, 7–8, 67–71
Fonda, Peter, 44
Foreign track, 65, 93–94
Futz, 31

Gardner, Arthur, 20
George, Ralph, 1
"Gettysburg," 22–23
Guns for San Sebastian, 32
"Gunsmoke," 39

Head mixer, 76
Herrenfeld, Kurt, 70
Horses, 42

Illusion, creating, 11–12
Independents, 9–10

Jazz Singer, The, 1
Job responsibilities, 2–3
Judgment, 22–23

Kansas City Bomber, 44
Kits, 26

Lassie, 8
Leader, 51
Levy-Gardner-Laven Production
    Company, 20
Library, 47–49, 83–90
    versus Foley stage, 71
Looping, 8–9, 20, 33–39
    editing, 38–39
    recording, 36–37

Magnetic tape, 9
Metro-Goldwyn-Mayer (MGM),
    1, 70
Minus dialog track, 65, 93–94
"Mission Impossible," 7
Mood, 12–14
Multiple effects, 45
Murmur, 42
Music and effects (M&E) track, 65,
    93–94
Music mixer, 76
Music versus sound effects, 100
Musifex track, 65, 93–94

Non-sync effects, 52–53

Off camera effects, 19–20
Outlaw Blues, 44, 49

Pre-dubbing, 65, 77–80
Preparation, 21–26, 80
Production sound, 32, 49
Production track, 27–32
Productivity, 9

Programming, 34–36
Projection room, 18

Reality, simulating, 11
Reynolds, Burt, 20

Saint James, Susan, 44
Schifrin, Lalo, 96
Separation, 71
Separation dubs, 93–94
Siegal, Don, 96
Silence, 14–15
Sinatra, Frank, 9
Slither, 73–74
Sound editing, 3–4
Sound editor, 2, 3
    and dubbing mixer, relationship
        between, 73–75
Sound effects
    art of, 4–5
    history of, 7–10
    multiple, 45
    purpose of, 11–15
    satisfaction with, 41–42
    selecting, 41–45
    stock, 48
    versus music, 100
Sound effects assistant, 3, 91–92
Sound effects companies,
        independent, 9–10
Sound effects library, 47–49, 83–90
    versus Foley stage, 71
Sound effects mixer, 76
Spartacus, 8
"Spiderman," 19
Spotting, 17–20
"Stagecoach," 99
Start mark, 18
Stock effect, 48
Sweetening, 45
Sync effects, 53–55

Tender Trap, The, 9
"Then Came Bronson," 83
Transfers, 47–48
Trim bin, 59

Universal Studios, 8, 48

Video system, 18

Walla, 12, 19, 42
Warner Brothers Studios, 1
Welch, Raquel, 44

White Lightning, 20
Wild lines, 37
Wild track, 49
Wiping, 30